PHOTOGRAPHIC MEMORY

Basic and Advanced Memory Techniques to Improve Your Memory - Mnemonic Techniques and Strategies to Enhance Memorization

EDOARDO
ZELONI MAGELLI

PHOTOGRAPHIC MEMORY

ISBN: 978-1-80111-961-0 - August 2019 - Original Version: Memoria Fotografica: Tecniche di Memoria di Base e Avanzate per Migliorare la Memoria - Tecniche Mnemoniche e Strategie per Migliorare la Memorizzazione

Author: Psychologist, Businessman and Consultant. Edoardo Zeloni Magelli, born in Prato in 1984. In 2010, soon after graduating in Psychology of Work and Organizations, he launched his first startup. As a Businessman he is CEO of Zeloni Corporation, a training company specialising in Business Applied Mental Sciences. His company is a reference point for anyone who wants to realize an idea or a project. As a scientist of the mind he is the father of Primordial Psychology and helps people to strengthen their minds in the shortest possible time. A music and sport-lover.

UPGRADE YOUR MIND → zelonimagelli.com

UPGRADE YOUR BUSINESS → zeloni.eu

CONTENTS

"Memory is the treasury and guardian of all things"

MARCUS TULLIUS CICERO

Introduction

Historians trace memory back to the days of Aristotle 2,000 years ago. In truth, it was Aristotle who first tried to understand memory when he stated that humans are born as a blank slate. This meant that everything we know, we only learned after being born. In ways, he was right as most of what we learn and remember happens during the course of our lifetime.

This book is not only meant to become a beginner's guide but also be seen as one of the most comprehensive books about improving your photographic memory. While most books in the market will look at either the basic or advanced techniques, *Photographic Memory* views the strategies of both. Furthermore, it will discuss the methods that you can use in your daily life to improve your memory with everyday tasks.

Chapter 1 is an introduction to your memory. You need to be able to understand what it is, how it works, and what parts it has before you can understand at least a piece of your memory. This chapter will discuss the memory process and what can interfere with it. Aside from that, you will be able to identify various types of memory before getting into the main one, which is a photographic memory.

Chapter 2 focuses on why you may want to improve your photographic memory. After all, if you are going to spend your time and energy learning all of the basic and advanced techniques related to it, you should know the benefits that come with enhancing your photographic memory. For example, what can it do for your academic performance?

Chapter 3 looks at the lifestyle changes that you may need to make in order to put the best effort into enhancing your memory. One of the topics I will discuss in this chapter is the importance of exercising and getting enough sleep for the mind. You will also look into how eating healthier foods and taking supplements will help you improve your brain function.

On top of this, you need to look at your stress level. What connects stress and memory, you may ask? Some people say that the former can be good for the latter, but many others believe that stress can affect your memory negatively, especially if it becomes chronic.

Chapter 4 will look at what people consider as the foundation or the most important technique of building your photographic memory: the *Memory Palace*. This is also known as your mind palace or the method of loci. If you have done previous research on the topic, you have probably run into similar terms pertaining to it. However, for the sake of this book, I will refer to it as the memory palace.

In this chapter, you will not only learn about the memory palace but also be able to set up your first memory palace as I take you through the steps. Then, you will manage to find out if you can have more than one memory palace.

Chapter 5 is going to discuss the *Mind's Eye*. Chances are, whether you have been researching about improving your memory or something else, you have learned about the mind's eye. However, when it comes to your memory, what does it mean? Furthermore, what important information do you need to know in order to make sure that your mind's eye is functioning correctly? After all, this is an important piece of your memory, so you have to guarantee that it is as clear as it can be. Otherwise, you may find yourself struggling. One specific aspect you will learn about is how observing and taking time to write down information will keep your mind's eye sharp.

Chapter 6 revolves around *Mind Mapping*. This is an important chapter because many beginners often get mixed up between the memory palace and mind mapping. While you will find similarities between the two, they also have a lot of differences. In this chapter, I will walk you through the proper way of creating your own mind map with necessary information.

You may find that you enjoy mind mapping more than creating a mind palace. However, both are extremely important for you to learn and practice as you are improving your memory.

Chapter 7 discusses *Mnemonics*. This is another important technique when it comes to improving your memory. Still, you won't just learn about how to perform a mnemonic. You are also going to learn the three fundamental principles that go into mnemonics, such as location, imagination, and association. You will also understand what types of mnemonics there are. Through this chapter, you should be able to find out which mnemonics are your favorite and which ones you will have to work on a bit more.

Chapter 8 is going to describe a variety of what many people consider to be some of the easiest memory techniques to use. Of course, it is important to know two factors when it comes to techniques which you consider easy. Firstly, most of the techniques will seem a bit hard at first. However, once you practice them a couple of times or so, you will start to realize how easy they all are. Secondly, the level of easiness from the beginning often depends on your personality. Just because someone says that *Memory Hooks* are one of the techniques that doesn't mean it will be for you. Therefore, you should not become discouraged if you feel that it is harder than one of the more advanced techniques in the following chapter.

Memorization is also going to be a focus for chapter 8. Other than learning about the *SEE Principle*, why writing down information is important, and the *Chunking Method*, you will receive tips on how to help you memorize information better. While not all of the techniques concentrate on memorization, most of them do. Because some people struggle with memorization, I have felt the need to include a few ways to help you reach your best success with memorization. A few methods we will discuss involve how often you should either listen to recordings or write information down.

Chapter 9 is going to focus on what some people call the more advanced techniques of improving your photographic memory. In this chapter, we will discuss the *Peg System, Car Method, Military Method*, as well as how to memorize a deck of cards.

We all struggle with remembering numbers and names from time to time. Therefore, chapter 10 is going to focus on some of the best methods to help us do that. For example, when it comes to names, you will learn that one of the most popular techniques is called the *Meeting Place Connection*. However, there are also two other connections, which are the *Character* and *Appearance Connections*. When you read about numbers, you will learn that you can use the *Number Shape Method* and the *Journey Technique*. You should also keep in mind that you have read about the chunking method in an earlier chapter. It is important to remember that the latter also works great when it comes to memorizing numbers.

Chapter 11 is going to not only give you tips in order to become successful at improving your memory but also help you learn about self-discipline. There are a variety of tips that you can use to upgrade your memory, such as staying focused and not allowing yourself to procrastinate.

Chapter 12 is the kind of seen-as-a-bonus section. It will offer you a couple of exercises so that you can start to practice a couple of techniques, if you haven't by the time you get to this chapter. However, one of the best parts about this chapter is it looks at a bonus method, which is called the *Emotional-Based Method*. While the majority of photographic memory techniques focus on memorization, there are a few which aim for emotion. This is important to focus on because emotion is one of the best ways that people will be able to encode, store, and recall the information within their memory bank. This bonus technique will describe a fictional story about a girl named Alessandra. You will read the story and write down the emotions you have through the story. At the same time, you will be able to pay attention to such things as facial expressions as you are meant to view this story inside your mind as you would if you were watching a movie.

Before we jump into what you need to learn about

your memory, it is important to remember that you will need to have patience when it comes to some of the techniques. You don't want to find yourself feeling burnt out as you try to learn every technique that is in the book as you are reading it. You never want to force yourself to learn the techniques to boost your memory as this is going to give you a negative view on how much work it takes to genuinely do that. In reality, enhancing your memory is one of the most beneficial steps that you can take when it comes to your mental health. Not only will you be able to remember things easier, but you will also be able to decrease your chances of acquiring cognitive diseases, such as dementia.

Keep in mind that you want to go slow and steady as you read this book. You don't have to learn the techniques as you read them. In truth, it is best to read and understand them before you decide to learn how to perform them. Doing this will help you find the most practical ways to start improving your memory.

Finally, it is important for you to know that your

learning does not stop here. You can continue to build your memory through my next two books in this series. The second one called *Memory Training* focuses on brain training and memory games. After that, you should check out the third book of the *Upgrade Your Memory* series, which is known as *Memory Improvement*. It completes the trio and concentrates on the healthy habits that you can install into your life in order to build your memory.

1. Get to Know Your Memory

Memories are one of our most important aspects of life. It helps us store information, gives us a sense of identity, and acts as a biography to our lives. Everything we know stays in our memory, which is stationed in our brain. We need it to perform tasks, as well as remember events, places, names, and job responsibilities. If it wasn't for our memory, we won't be able to communicate, know the names of animals, friends, or family, and even complete daily tasks.

We all know something about memory. We understand what it does and how important it is. We know that it is an extremely complex system, which scientists have studied for decades.

Their ultimate goal is to figure out how and why it works the way it does.

The Memory Process

The memory process has three parts.

Encoding

Encoding is the first stage in terms of processing memories. At this point, the information starts heading into our memory, so we will be able to remember it later. If it is not encoded, we will not have a recollection of it. Because the information comes from our sensory input, it changes into a form that encoding can work with. For example, while we will see a word in a book, our memory will encode it through sound, visual, or meaning. These are the only three ways in which encoding occurs.

When we encode new information into our memory, we connect it to someone we already know. Say, if you need to remember 3121, you may sing the numbers to yourself because of the way they sound together. You may also find meaning within the list of numbers or remember it as a visual. No matter how you think of these digits, you will be able to connect 3121 to something that you already know.

There are other ways that our brains encode data. The first one is through automatic processing. This means that we aren't even aware of what we are doing. It doesn't take any effort from us at all. The examples of automatic processing are details like time and dates. Furthermore, there's effortful processing, which occurs when we are trying to remember important events, such as studying for an exam.

Storage

Storage is the second stage of the memory process, which speaks of how long we hold information over time. There are several factors that will influence how many days or years a detail can remain in our brains. For one, it depends on which area of our memory storage information can be found. The only options are short-term memory, long-term memory, and sensory memory.

When information is placed in our short-term memory, it comes from sensory memory. This type is limited to a certain amount of time. We usually only hold information in short-term memory for about a minute. You are using short-term memory when you are trying to remember a message so that you can quickly write it down. There is a restricted amount of space in our short-term memory as it only holds about seven pieces of information on average.

Meanwhile, there is no cap when it comes to long-term memory. We can hold information in this area for the rest of our lives. However, this doesn't mean that we will be able to retrieve the data for as long as we want. How you retrieve information depends on the method that you have used while processing it.

Sensory memory will hold a lot of detailed information but only for about a second. The data will then either move onto short-term memory or remain unprocessed.

The other factors that influence time include our age, any memory problems, allure of the details, how we encode the information, and the data's level of importance.

Retrieval

Retrieval is the third step of memory processing, and it occurs when you bring the information out of storage. Trying to retrieve ideas will allow us to know if it is within our short-term and long-term memory. If the information is part of the former, we will be able to retrieve it the same way we have stored it. For example, if we remembered a list of numbers in a certain order — say, 21314151 — we would recall it exactly like that. When information is retrieved from our long-term memory, on the other hand, it is done through association. You can think of something because of its connection with an image or emotion.

There are a lot of factors that can affect the retrieval stage, such as what other information you have stored since and how you have kept that memory. If you are trying to remember an event from five years ago, for instance, you will have a tougher time retrieving the information than something that you have kept in your mind five months ago. You will also be able to recall an event easier if you use certain cues, such as sound or image. There are three main retrieval types.

1. Free Recall

This happens when people can remember the information in any order. This type has two effects, namely the recency effect and primacy effect. The former takes place when a person thinks of something at the end of the list more than what's in the beginning. The opposite of this is the primacy effect in which starting items are easier to remember than the ones at the end of the list.

2. Serial Recall

Primary and recency effects are also a part of series recall. It occurs when you remember events in the order in which they have happened. For example, if you are going for your morning walk and see a man walking his dog, a group of kids running through a sprinkler, and a woman carrying groceries in their house, you will have a recollection of such activities in that exact order. You will probably recall the information through a series of images you have encoded in your memory.

3. Cued Recall

The cued recall takes place when you process information along with cues. There have been many psychological studies to prove that people who use cued recall remember information better if the link between the information and the cue is stronger. We often used it when we are looking for information, which has been lost within our memory.

Interference With the Memory Process

The memory process does not always occur as smoothly as we hope. In fact, there are various amounts of interference that can take place when we are trying to process and retrieve them.

1. Retroactive Interference

Radioactive interference happens when you learn something new right after previously getting a different information previously. We commonly experience it in a classroom as we spend 50 minutes to learn about the lesson for the day. We start by feeling that we will be able to remember everything we are taught. However, by the time the class ends, we don't retain much of what we have heard in the beginning. The reason is that as we continue to learn new stuff, the newer ones can interfere with the older information, especially if they come to you at close intervals.

2. Proactive Interference

Proactive interference occurs when you are having trouble gaining new information because of the things that are already installed in your long-term memory. It often happens when the information you are trying to store is similar to what you have previously learned. For example, you are trying to remember your new address, but you are struggling because your brain is more accustomed to the old one.

3. Retrieval Failure

Retrieval failure takes place because the information has begun to decay within your memory. It is similar to when you struggle to remember how to prepare a meal that you haven't cooked in years or perform an algebraic problem.

It is important to note that some people believe that there are four stages of memory processing, not just three. While most agree with encoding, storage, and retrieval as the official steps, others say the first stage is attention ("Types of Memory", n.d.).

The information you are going to encode supposedly needs to gain your attention first. If it has not gone through this phase, we may not be able to remember a lot of things. Think about the last time you gave heard something interesting vs. something uninteresting. You are more likely to recall the former since it has "caught your attention" than the latter.

Types of Memory

You already know a few types of memory, e.g., short-term, sensory, and long-term. However, they have subtypes that you should learn about as well..

Sensory Memory

Sensory memory is attached to the five senses of sight, hearing, taste, smell, and touch. Therefore, its subtypes are related to at least one of your senses.

1. Iconic Memory

Iconic memory is a part of your visuals. It is attached to your sight, such as seeing bright colors with a dark background. Through this subtype, the colors will be encoded into your memory. Thus, you can remember the shape and colors of certain objects but perhaps not the background. The iconic memory allows us to remember things or images seen even for a few moments.

2. Haptic Memory

Haptic memory usually only lasts for a few seconds. It responds from what we feel, such as a pinch, hug, etc. When we feel that something is cold, for instance, this is our haptic memory putting forth its effort to instill in your brain that ice is cold.

3. Echoic Memory

When our memory is trying to convert what we have just heard into our short-term memory, it is using echoic memory. The latter is at work when your mind replays information as you try to remember a message that you want to write down. It only takes three to four seconds before the idea moves into your short-term memory.

Many people feel that there are two other subtypes of sensory memory that correlate with our sense of smell and taste. The problem is, they have not been studied yet. Furthermore, scientists have only recently begun to study iconic, haptic, and echoic memories. While this means that there is little known about the subtypes mentioned above, we do know that what starts with our sensory memory usually transfers into our short-term memory.

Short-Term Memory

Short-term memory includes working memory. While they are similar as they hold information for a brief period, there are also differences between the two.

Short-term memory will often use techniques — say, *chunking* — that allows you to hold more information than usual. Instead of remembering seven names, for instance, you will be able to remember 14 names because you can group them together. Working memory, meanwhile, is the part of short-term memory that holds information through an auditory or visual-looping process. This means that the information will continuously play on repeat, so you won't forget it quickly. The information within working memory is often manipulated, which makes it easier to remember for some time.

There are three phases within working memory. The first one is the *Phonological Loop*, which we have just discussed. The second stage is the *Visuospatial Sketchpad*, which usually works with the first phase. For example, if you need to remember a seven-digit phone number, you will remember it better if you not only repeat it — phonological looping — but also use visuals, which is the visuospatial sketchpad.

The third one is the *Central Executive Phase*, which combines the phonological loop and visuospatial sketchpad into one. At this point, the working memory is connected to the long-term memory, considering the central executive will transfer the information into the latter.

Long-Term Memory

If you want to remember what you need to do tomorrow, you have to store this information in your long-term memory today. This is the only type of memory that will hold on to what you have learned forever. Now, long-term memory has two main subtypes.

1. Implicit Memory

People often refer to implicit memory as unconscious memory. This type refers to the activity that we learn over time. For example, when we are trying to build our skills, we are using our implicit memory. It also works when we start to do something without thinking about it, such as typing on a keyboard without having to look at the keys, tying our shoelaces, and washing the dishes.

2. Explicit Memory

Explicit memory is commonly known as conscious memory. This is the form of memory that we use when we are thinking about our actions. Essentially, it is the opposite of implicit memory. This subtype, nonetheless, is divided into two parts.

The first division is the *Episodic Memory*, which focuses on the specific moments that you remember. For example, you might recall spending the Fourth of July with your grandparents when you were younger. You might also vividly remember parts of the event, such as standing in the back of your grandfather's red pick-up truck to watch the fireworks, eating on a white picnic table, and seeing your grandparent's farm. In general, you have a recollection of the what, where, when, and who, which are all related to a particular occasion. Another example of explicit or flashbulb memories (as some people may call it) involves remembering exactly where you were when you heard Martin Luther King Jr. had been shot or when the September 11, 2001 attacks took place.

The second division is *Semantic Memory*, which refers to the retrieval of factual information. The latter typically come from schoolbooks, places, or concepts that have heard about or seen before. The facts of life that we have learned over time are encoded into this type of memory as well. Say, you can remember what to do once you go to the grocery store. You know that you are supposed to pick up the items you need, pay for them, and leave the store.

Photographic Memory

One type of memory that people don't often discuss is the photographic memory. Imagine being able to remember a person, place, or object simply because you have an image of it in your mind and describe it in detail. You can recall the design on your friend's Double Excess T-shirt, the main words you read on a page within a book, or even the songs on the DJ's list in order.

Eidetic Memory is often another name for

photographic memory. However, there is one distinction between the two. You are talking about the former when you remember a visual after turning away from it. You have probably stared at an object, such as a vase, for a couple of seconds and then looked the other way later. If you still see that vase within your mind and remember its colors and design, this is your eidetic memory at work. Its main distinction with photographic memory, however, is that the image remains in your memory for only a few seconds. When you have a photographic memory, you can remember things for a long period of time as it is stored in your long-term memory and not in your sensory or short-term memory, which is where the eidetic memory lies (Beasley, 2018).

Distinguishing the two is important to keep in mind throughout this book, as well as if you continue to do your own research on photographic memory. Several sources will use eidetic and photographic memories interchangeably, which can easily become confusing to people. However, as long as you remember their differences, you will be able to upgrade your memory with ease. While some individuals have stronger photographic memories than others, it isn't because they were born with a special gift. The more realistic reason is that they use different techniques to strengthen their ability to remember things.

2. Benefits of Photographic Memory

Why should you be interested in learning about photographic memory? After all, it isn't exactly what you probably think it is. You may feel that you already have a pretty good memory as well.

One factor to note — other than the variety of benefits we will discuss in this chapter — is that memory decays. The older we get, the more we will struggle to remember our childhood memories, what we need to pick up at the grocery store, why we have come into a certain room, etc. Among the biggest benefits of building your photographic memory is that you will learn dozens of techniques to engage your memory.

This will make your brain more energetic and capable of holding more information. Not to mention, it can slow down the natural decaying process that our memory database may experience.

You Will Perform Better Academically

One of the downfalls of trying to perform well on a college exam is that you have so much information to remember. However, the truth is that we often struggle with memorization because we are too focused on words and definitions. How many times have you used index cards to try to recall what a certain word means? This is usually a technique that people use when it comes to memorization. Nevertheless, there are many other techniques used to improve your photographic memory that will make this task easier for you.

In reality, photographic memory has helped so many people perform better in school; that's why another name for it is "encyclopedia memory" ("The Good and Bad Things," n.d.). The reason is that the individuals who study using the strategies that may improve their photographic memory are able to remember the details that other students don't.

Furthermore, photographic memory will help you learn different techniques to recall what you are learning and keep it in your memory bank longer than ever. If you are or have been a college student, you understand how fast-paced your classes can be, especially in the summer. Sometimes, you have to study a whole chapter or two out of a thick textbook within one period.

Photographic memory will help you learn more in less time. When you strengthen your photographic memory, though, you are not just looking at images but also focusing on what you hear. This trait is especially important when you need to highlight information or write or type your notes.

You Will Remember More Information in Detail

When it comes to photographic memory, it doesn't matter whether you are trying to think of an image or a series of numbers or words. What matters here is the strategies that can help you remember them.

The important factor is making sure that you have a strong photographic memory. The stronger your photographic memory is, the more information and visuals you will be able to store in your mind. Think of how many times you have tried to recall a detail that you have seen in a photograph, but then a few minutes later, you realize that you have no idea where the lamp is, what color a person's shirt is, or where the window is located.

With a photographic memory, though, you will be able to remember all these details easily for a longer period.

Photographic Memory Boosts Your Confidence

How do you feel when you don't remember the information that you used to know? How do you feel when you forget someone's name or what their interests are? Think back to the time when you studied for a test, but when you had to take it, you could not recall much of what you learned. Similarly, when you go to the grocery store without your list, you may struggle to remember what you need to buy. There are a lot of features about life that we tend to forget, including needing to purchase treats that our children can bring to school or telling them that you won't be home till their bedtime.

Just like everyone else, you have forgotten something important in your life, which has made you feel sad, frustrated, or even angry. While you try to tell yourself that it happens and try to move on, there is always a part of yourself that holds on to your forgetful nature as you find yourself forgetting more and more things. Sometimes, you may even wonder if there is something wrong with you.

Well, I will tell you right now that there is nothing wrong with you. It is common to fail to call to mind various details of our life throughout your day, regardless of how important they may be. It can be due to stress, lack of sleep, having too much to remember, as well as not having an organized system for this. Another reason is that you don't have a strong photographic memory.

Because you can only recollect vital aspects of your life with a dependable photographic memory, it will help you boost your confidence. You will start feeling like you can remember what you need to tell your children or pick up at the store. You may also feel like you can become organized so that you can think of everything you have to do without stressing about the messy details or letting it keep you up when you are trying to sleep.

You Will Become More Mindful

We often get involved with a task or start thinking endlessly about it and don't pay attention to what we are doing. This is called mindlessness, and it can cause a lot of problems within our lives. A common example of mindlessness is when you are driving to work and don't remember passing certain landmarks. E.g., a small lake or town.

On the other hand, you can claim that you are mindful when you exhibit awareness towards your surroundings. After all, you know what you are doing, and you remember your actions.

When you improve your memory, you need to become more aware of the information that you want to retain. You should start to pay more attention to your environment, as well as what you are reading, feeling, and hearing. As you show more consciousness, you will become more mindful of everything you do. Even when you don't need to hold the event, you will still know what you are doing and why instead of working on things aimlessly.

Becoming mindful can help you lead a healthier life. You will become more conscious of what and how much you are eating, as well as when you feel full. You can also be more aware of how much sleep you are getting and what thoughts come to mind. In return, it can further boost your self-esteem and lead to greater success because you will be able to focus more on positive ideas.

You Will Become a More Compelling Public Speaker

Many of us have jobs that require us to speak in front of people. For instance, you may have to present a new product or idea in front of a committee, train new employees, or work in customer service and always have strangers to talk to. No matter what your line of work may be, communicating with dozens of people can be difficult, especially when you need to be persuasive.

If you have ever spoken before several individuals within a room, you know that you need to retain eye contact as much as possible. This means that you don't want to hold the paper that's filled with your notes, look down at it often, and talk to your paper. If you struggle with public speaking or can't seem to remember your speech, you are going to struggle with eye contact.

One benefit of improving your memory is that you will be able to memorize your notes better. You can study and understand your speech, so you don't have to spend a lot of time looking at your paper to make sure that you are saying everything. You don't have to worry about getting lost within your paper and stumbling over words while you are trying to find your place as well. Instead, you can go up in front of a group of people and talk with confidence as you can recall the main points of your speech. It will allow you to remember the rest of it, for sure.

Now, the suggestion above is no indication that you should not have a paper with your notes in front of you. Most speakers have some type of notes in their hands, to be honest. However, you have to avoid using them too much to be able to keep eye contact with your audience and be more persuasive.

You Will Have Deeper Relationships

People enjoy the company of others who remember something about them. This makes them feel like you care about them. You spend your time trying to call their favorite food or movies, how many children they have, if they have any pets, what their occupation is, and so much more. Furthermore, you will feel more connected to them because you can remember certain information that the other folks may not know about them. This can help in any relationship, whether it is with your significant other, friend, relative, or co-worker.

You Will Become More Productive

As you start to improve your memory, you may feel yourself becoming more productive. While part of this is because your confidence is increasing, the other reason is that you use less of your energy trying to remember some information. When we dig into our memory database, we use some of our daily energy. This causes us to feel tired, and we become less focused since we are also losing our interest and productivity in the process.

Think of how you start to feel near the end of your workday compared to what you have felt at the beginning of your shift. When you go to work, you feel more energized because your body and mind still feel well-rested. You feel like you are ready to take on the day and accomplish all your tasks. However, as the day goes by, you start to slow down and notice that you are becoming more tired. This is because you have used a lot of your daily energy to try to remember what you need to do, how to do it, and how to solve a problem.

The more you improve your photographic memory, the easier it can be to remember some information for your tasks. Thus, when the end of the day comes, you will still feel like you can take on the world.

Other Benefits

There are dozens of benefits when it comes to improving your memory. While I cannot discuss them all in this book, here is a list of perks that you will receive once you upgrade your photographic memory.

- You can remember shopping lists better, which will make you less likely to forget any item.

- You can recall someone's name.

- You will be able to remember an address easier than ever.

- You can remember all the tasks you need to accomplish for that day.

- You will be able to handle calculations more easily.

- You can recall phone, account, PIN, and other sequences of numbers better.

- You will be able to learn a foreign language easier as you will gain a better understanding of their terms and pronunciations.

- You will remember directions easier.

3. Lifestyle Improvements for Your Memory

If you know you have lifestyle habits that you can improve on, you are more likely to improve your memory. It takes a lot of energy for your body to function throughout the day, you see. Because of this, you need to make sure you eat well, get enough sleep, and take on other healthy habits.

This chapter isn't about making sure that you live the best and healthiest life possible. It is about how your well-being affects your memory. This means that the better you feel overall, the more your memory will improve. Some of the lifestyle improvements discussed below may already be familiar to you, which is great. These are the common steps that people can make in order to boost their memory.

Exercise

Exercise is not always something that we want to do, but it is necessary for our overall health. As we work out, we start to feel better mentally and physically. This helps improve our memory and decreases the risk of dementia.

Several studies prove the importance of exercise to brain health. Not only have results shown that secretion of neuroprotective proteins increases, but the development of neuron also improves. Furthermore, a study whose correspondents revolve around the ages of 19 and 93 improved their memory performance by spending 15 to 20 minutes on a stationary bicycle (Kubala, 2018).

Get Enough Sleep

Just like exercise, sleep is also important when it comes to our memory. As I have discussed briefly before, the more alert you are throughout the day, the more energy you have to put towards your memories. A good sleep keeps your psycho-emotional balance and of course, with low levels of anxiety and stress you'll be able to remember better.

Sleeping well is very important for the enhancement of cognitive functions, such as learning, attention and concentration. Sleep is essential for cognitive performance and plays a key role in the memorization process. While we sleep, mnestic traces are enhanced and reactivated and incorporated into the long-term memory database.

One of the biggest reasons sleep disturbances disturb the memory function is because it hinders the transfer of memories from the short-term memory database to the long-term memory database.. When you get the sleep needed, it triggers the parts of the brain that connects the process with brain cells. Therefore, the more sleep you get, the easier the transference will become ("Improve Your Memory With a Good Night's Sleep," n.d.). REM sleep is essential for memory consolidation. It has been shown that without REM sleep memory does not consolidate.

Furthermore, our brain is still active when we are sleeping. While we are resting, it connects the information that we have learned from our previous or older memories. It often gives us dreams or reasons to have "aha!" moments the following day. It can allow us to solve problems that we have been struggling with earlier.

Eat Healthily

One way to improve your brain function is to eat healthily or follow a "memory diet." One such diet can be the Mediterranean Diet as it is known to boost memory and slow down cognitive decline due to age. It mostly consists of fruits, seasonal vegetables, whole wheat grains, herbs, nuts, legumes and extra virgin olive oil col pressed. You will also eat more fish and seafood than red or lean meat. However, you want to eat more chicken or turkey than beef and other red meat.

If you are a senior citizen, it is best to look at the MIND diet, which stands for Mediterranean-DASH Intervention for Neurodegenerative Delay and is similar to the Mediterranean diet. In truth, studies have shown that this diet helped reduce signs of Alzheimer's disease by 53% (Alban, 2018). However, you need to get at least three servings of whole grains a day and an ounce of nuts. You should also have a salad and another veggie dish every day, as well as chicken and berries twice a week. The foods that you need to have more than once a week include fish and legumes.

Take Supplements

If you are like everyone else, you probably lead a busy life. In fact, you may feel that you just don't have the time to make sure that you can follow a specific diet right now. If you can relate to that, many people say that you should try taking memory supplements, such as fish oil, multivitamin, and curcumin.

It is important to note that the pills should not replace the amount of sleep or exercise that you need every day. You still should eat healthy foods as much as possible as well.

Watch the Amount of Stress That You Deal With

Dealing with a bit of stress is fine for your memory. In truth, acute stress can even boost it. However, having a large amount of chronic stress is going to cause memory loss.

You may have noticed this with yourself when you feel too stressed out. You find yourself forgetting about going to your children's doctor appointments, attending business meetings, returning the library books in time, as well as doing other errands that you need to accomplish within the day.

Most people will start to worry about their memory loss and fear that it is an early sign of Alzheimer's Disease or another condition. However, while it is always a good idea to get checked by your doctor, chances are, you are merely being controlled by chronic stress.

For example, Maria is a 33-year-old mother of three kids whose ages are between 2 and 7. She and her husband work two jobs each so that they can support their family, live comfortably, save for their children's college education, and prepare for retirement. Maria is constantly under a lot of chronic stress between working from 60 to 70 hours a week, cleaning, taking care of the children, cooking, making sure bills are paid, and running other errands. Lately, she has noticed that she forgets to pay bills on time, bring her kids to their appointments, transfer money into the right accounts, and buy essential supplies at the grocery store.

Because Maria is afraid of what is happening, she makes an appointment with her primary healthcare physician. This doctor informs Maria that the only problem is that she is handling a lot of stressful things at once. In order to improve her memory, one of the first steps she needs to take is to let go of some of them.

After talking to her husband, they decide that Maria will leave her part-time job, which will give her 20 to 30 hours a week to take care of the family and house instead. Since then, Maria has noticed that she can remember to run all of her errands again, pay their bills on time, and make sure that their children get to their appointments.

Other Ways to Improve Your Memory

- Limit your alcohol use

- Stop smoking

- Meditate

- Keep your mind stimulated

- Get fresh air

- Maintain a positive mindset

- Get out and enjoy your life

4. Memory Palace

The memory palace is also known as a *Method of Loci* or *Mind Palace* (Loci is the plural of the Latin term locus, meaning "place"). This concept has been around since ancient Rome and is essential to understand when you are working towards improving your photographic memory. The memory palace is an imaginary spot in your mind that you have based on a real location.

For example, you know what your bedroom looks

like without having to be there. You can also describe your office at work even if you are not inside it. After all, you can use the mental pictures in your brain to connect it to what you need to remember.

How Does a Memory Palace Work?

When you think of a memory palace, you should think of a home construction to understand how it works. You can build the rooms in your house one by one as you need to remember other tasks, such as buying things to fill them and setting up other areas that you need to complete that week. With each list, you get to build a new room in your memory palace. Each time you build a room or add information to an existing one, you continue to strengthen your memory palace. These details will be stored in your palace, and you will be able to recall them at any moment.

Setting Up Your Own Memory Palace

In order to further explain how you should set up your memory palace, let's walk through a few tips together.

1. Pick a familiar place

You can pick any place in this case, but you want to make sure that you remember everything about it. For example, if you choose your living room, you should be able to recall its shape or where you have placed different types of furniture. If you decide to go with your office, you want to do the same thing. It is always a good thing to look around your chosen room before you continue to make sure that you will miss anything that can be essential for your memory palace. While we know the places we see every day, we can forget about certain objects because they are always there. We just don't think about them very often, so you may not remember their placement when you are trying to create your mind palace.

Once it is time to start recalling your list, you need to imagine heading to your location of choice. If you pick your living room, for instance, you want to picture yourself walking up to and in your house and then entering your living room. You can also imagine yourself walking from your bedroom, into your hallway and then the living room. You don't want to create a specific scene in this step — just visualize yourself going to your chosen location.

2. Make a list of what you remember

As you are walking to your living room, you want to remember all the items that you see while doing it. For example, if you come from the bedroom and go to the living room, you will imagine walking out of your bedroom and turning down the hall towards your living room. You can also picture out the doorway that leads to other rooms, any photos that may be hanging on the walls, as well as end tables or pieces of furniture that are in the hallway. Similarly, you can imagine parts of the living room that you can see from the hallway, such as a plant aquarium or clock on the wall.

3. Designate and associate

This bit sometimes becomes tricky for people, but many others have fun with it. When you need to start designating and associating stuff, it means that you have to choose the items you imagine around your location and connect them to what's on your list. The thing is, you want to create an image in your mind that you are going to remember. You want it to stand out. and the best way to do this is by turning your everyday item into something interesting and crazy. The crazier it is, the better!

For example, when you notice a doorway in your hallway, you may think that it is made out of yellow Post-it notes, just like the ones in your shopping list. You may imagine the end table in the hallway as a head of cauliflower because you need to pick up cauliflower at the store. You may also picture out the fish swimming in blueberry juice on one side and aloe vera juice on the other side. You will want to associate each item on your list with an item that you have seen in your location.

A specific trick that many people don't think of right away is to associate the things that they need to pick up in chronological order. For example, if you are going to the center of your city because you need household items and groceries, you will pick the former before the latter. Therefore, you want to make sure to imagine all your household items, preferably in the way you will pick them up at the store, at the beginning of your location before moving on to the groceries. When it comes to recalling your list, it will help to recall the item in the same order that you will place them into your cart.

You should always bear in mind that practice makes perfect. It is always a good idea, especially as you are getting used to your memory palace, to write the list down in the same order that you will pick up the items at the store. Then, take the list with you when you go shopping. However, do not look at it unless you are having trouble recalling some things or need to double-check it to make sure that you have picked up everything before checking out.

You Can Have More Than One Memory Palace

Many people often wonder if they are allowed to have more than one memory palace. The truth is, you can. When you are starting to build your mind palace, though, it is best to stick to one for a while or at least until you become comfortable transferring from one memory palace to the other.

In fact, once you are 100% comfortable with your first memory palace, you may then think of creating a second one and then a third, fourth, and so on. There is no limit when it comes to how many mind palaces you can make as long as you are comfortable with the number and can still continue to jump from one to the next.

How does transferring from one memory palace to

another work, you may ask? It basically depends on your list. Each list that you establish in your memory palace is going to remain there, especially if you recall the list every now and then. That said, you cannot help but lose track of some lists. For example, you may forget your grocery list as they tend to change every week. However, you can always recall the other sets that you want to keep in your memory back, such as the names of 45 flowers or 45 Presidents of the United States.

It is important to note that both of the lists mentioned above will have their own memory palace. For instance, you will start by associating the 45 presidents to objects within your office. Then, once you have accomplished and practiced it, and you find no trouble with this memory palace, you will be able to move on to the next list. Each flower can become associated with a president, too. Say, George Washington is likened to a red rose, John Adams seems like a sunflower to you, and Thomas Jefferson can become a lilac. But this is another technique.

5. Mind's Eye

You will get to know your *Mind's Eye* better than ever as you improve your photographic memory. This is because your mind's eye is a part of your mind that allows you to remember rooms, objects, or anything else exactly as they are.

Its definition is to be able to think of what is not directly in front of us (Friedersdorf, 2014). However, your mind's eye can do more than allow you to see what you know even when it isn't there. In reality, it is also capable of creating special images for you. For example, if someone tells you to imagine a purple cat with a black witch's hat swinging on the power lines, you will be able to picture that out perfectly.

One of the best tips when it comes to using your mind's eye is to do what you can to limit your distractions. It is going to build an image through your five senses, you see. Therefore, when you are distracted, you won't be able to pay attention to what you hear, smell, feel, taste, or see. This can cause interruptions with your mind's eye and make it harder to create images that you can recall later.

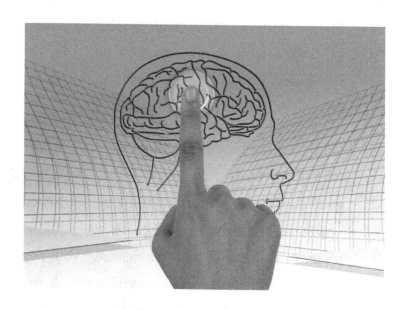

Keeping Your Mind's Eye Clear

Everyone struggles with keeping distractions away

from time to time. Therefore, there are a lot of techniques that you can use in order to keep your mind's eye clear from getting disturbed.

Observation Is Key

Some people are natural when it comes to being observant, but others struggle with it. If you find that you are more of the latter, you want to build your observation skills as they are important in developing your mind's eye. The best way to do this is by observing items around your home and outside. You can start by looking carefully at a vase placed in your living room. Notice the colors and the designs on the vase. You don't need to touch or pick up the vase, simply stand in front of the vase and observe everything. You may notice a chip on the top or how a part of the paint is starting to chip away. Note all this information and then leave the room. You will then try to remember as much about the vase as you can with your mind's eye. After you are able to imagine that, you should go back and see how well you have remembered all the details.

You can further test your observation skills by leaving the room and waiting for a couple of minutes before trying to picture out the vase. You can then either draw it or go back into the room to see how close you have come to remember every detail of the vase.

Write the Information Down

When you start to observe items, nature, or features of a room, you will find yourself becoming distracted. You will notice your mind wandering to something that you are not supposed to check out. When this happens, one of the best techniques is to start writing down what you are observing. For example, you are sitting outside on your porch and trying to look at the large tree in your neighbor's front yard. However, you struggle to keep your eye on it because you have taken a glance at their house and become distracted by people walking down the street, dogs barking, and kids playing. To avoid forgetting what you are doing, you should write down everything that you have observed about the tree. For starters, focus on the trunk of the tree. You will notice how the bark moves up the tree, how some of it is missing, and then you start to see where the branches begin. You need to describe the branches and the leaves on the paper, ending it with how the tree is taller than the house.

Stop and Smell the Roses

We have all heard the expression that we sometimes need to "stop and smell the roses." This means that you are moving too fast in life and not enjoying some of its best features. Perhaps you aren't spending quality time with your family, you don't allow yourself to take in the beauty of nature, or you don't stop and look at your surroundings. Whatever the case is, you want to take the time to observe what is around you randomly throughout your day to be able to cherish what you have.

Many busy people who struggle with handling their

stress find this to be one of the best ways to recognize how blessed they are. When they start to feel overwhelmed, they will stop what they are doing whenever possible and check out their environment. They will notice the people around them, what they are doing, as well as how their voices sound. They will see the bugs on the flowers or the birds flying in the sky. You don't need to observe your surroundings for a long period of time; you just have to make sure that you have at least a few minutes to observe where you are and what is going on around you. Not only will this boost your observation skills, but it will also help you connect to the world.

Part of enhancing your photographic memory is learning as much as you can so that you can associate certain items to the things you need to remember. The more knowledge you hold, the easier association becomes for you.

6. Mind Mapping

Science has repeatedly shown that the brain contains an enormous potential that is only waiting to be released. One of the ways to unlock this potential is to start using the mind-mapping method of Tony Buzan and Barry Buzan (2018).

This powerful tool, in addition to exploiting your innate potential, helps you to organize your thoughts, think better and above all to remember what you learn. Mind maps use fundamental elements for the overall functioning of the brain, such as: visual rhythm, schematizations, colors, images, imagination, different dimensions, spatial awareness, Gestalt and tendency to complete associations. This system allows you to use the full range of your mental abilities. It will help you in creativity, problem solving, planning, memory, thinking and face the changes.

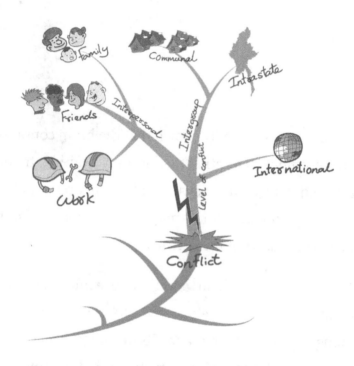

I would like to open a parenthesis on the great Leonardo da Vinci, not only because he was born a few miles from my house, but because like other great geniuses of the past, he managed to draw on a wider range of mental abilities than their peers. In fact, the great minds of the past have used a much larger portion of the mental capacities that each of us has. What makes Leonardo's mind special? His brain instead of thinking in a more linear way than his contemporaries, began to intuitively use the principles of mind maps, and therefore of *Radiant Thinking*.

This form of thinking is the simplest and most natural way to use the brain because in fact, our brain already contains mental maps.

The thinking mechanism of the brain is like a sophisticated device capable of producing branched associations, with lines of thought that radiate to an infinite number of information and data. This structure reflects the neural networks that reflect the physical architecture of the brain.

If we analyze Leonardo's notes, we can see words, symbols, sequences, lists, analyses, associations, visual rhythm, Gestalt, different dimensions, numbers and figures. This is an example of a complete mind that expresses itself globally and makes a whole use of its cortical activities.

It will be difficult to match Leonardo's genius, but surely this powerful tool helps us to release the immense potential that we have in our brain. Try, you will be satisfied with your mental performance.

Mind Map Essentials

Why mind maps help us learn and remember better than traditional notes? First of all, traditional notes are monochrome and monotonous. The notes of a single color are difficult to remember, they are boring and therefore they will be forgotten because the brain gets bored, turns off and tends to ignore them. They are predisposed to put the brain to sleep. It is a methodology that does not exploit the capabilities of our cerebral cortex and this limits the capacities associated with our left and right hemispheres. Therefore, these abilities cannot interact with each other and hinder a virtuous circle of movement and growth. This linear writing of the notes encourages us to refuse learning and to forget what we have learned. It prevents the brain from making associations, limiting your creativity and your memory. It is a mental narcotic that slows down and inhibits your thinking processes.

Instead, creating mind maps allows you to work with keywords that immediately convey important ideas and concepts, obscuring a long series of words that have lesser importance. This allows your brain to make appropriate associations between key concepts.

If you want to take notes effectively, there are 3 fundamental things you need to remember: *Brevity*, *Efficiency* and *Active Involvement*.

This is why mind mapping is known as one of the best methods to encode and retrieve information from your memory database. While every list you create through mind mapping will be different, all minds are organized in a specific way, which makes them similar. They all use imagination to easily remember things, as well as colors that make items stand out. When you think of a mind map, you want to think of a regular city map or the map at a mall. There is always the center and then everything else branches out from there.

When it comes to mind mapping, there are five aspects that you need to have.

1. You need to have a center. This will be your main subject or idea, such as the Cold War.

2. Each theme that comes from your center will be comprised as branches. For example, one branch of the Cold War refers to why it has happened, another one is the Berlin Wall, and the next one consists of the aftermath.

3. Each branch has a keyword or image that you can associate to your memory bank. For example, with the Berlin Wall, you can picture out a wall.

4. You can also create twigs with less-important themes that come off from your main branches. This is just like a branch of a tree that has smaller twigs or branches attached to it. The trick is to make sure that each twig is relevant to its branch.

5. A nodal structure will form through the branches.

Create Your Mind Map

You can use any type of idea or theme to create your mind map.

First, you want to start at the center, which is the main idea of your mind map. You can create an image as a part of your idea or use a keyword. Whatever you do, you want to make it colorful, something that you can easily remember. Therefore, it will help to make your image a little cartoon-like, crazy, and vibrant.

Second, you want to get your branch themes, which flow from the center image. In order to help yourself with this process, you can brainstorm and write the branch themes down ahead of time. You can also do this with any subtheme, which you will add in later. For example, if your central theme is food, your branches can consist of meat, fish, vegetables and whole grains. You can recall it better by creating an image with each branch, turning the branch into a different color, or just using a keyword.

Third, you should add the subtopics or your twigs. Just like what you have with the branches, you can make them as colorful and fun as you want.

It is important to realize that a map never really ends. You can create as many subtopics as you wish. All you have to do is to relate to the branch's theme with the central idea. In fact, you will probably find yourself adding information to your mind map as you continue to gather more details about the topic.

The theme of mind maps would deserve a separate book, if you want to learn how to master this powerful technique, I suggest you study the book "The Mind Map Book" by Tony Buzan and Barry Buzan

7. The Family of Mnemonics

You use mnemonics often to remember certain information. For example, "Naughty Elephants Spray Water" is a mnemonic for directions, namely North, South, East, and West.

Schools often use similar phrases to teach directions to children. Mnemonics can take several forms, such as song lyrics, rhymes, expressions, models, connections, and acronyms.

Fundamental Principles of Mnemonics

Before we get too far into detail regarding the various forms of mnemonics, we need to discuss its foundation. There are three fundamental points: *association, location* and *imagination.*

Association

Association takes place when you link what you want to remember to what you are going to remember it by. For example, when you think that Thomas Jefferson was the third President of the United States and the author of the Declaration of Independence, you can picture out the Declaration of Independence or the number 3 shaped as Thomas Jefferson. It is important to note that when you are coming up with your own associations, you need to figure them out yourself. You will be able to remember this information better if you associate it with something that you have thought of.

There are many ways to remember things by association. Other than with images and numbers, you can merge the objects together, place one on top of the other, or imagine the two objects dancing around together or wrapped around one another. You want to let your mind get as creative as it can be. Remember, this is not the type of information that you will need to share with anyone else. Therefore, you don't have to worry about what others will think of your associations. What matters is that you are able to retrieve them from your memory database quickly.

Location

When you focus on the location, you are giving

yourself two things: separating one mnemonic from the other and providing context that allows you to place the mnemonics together. This way you will be able to separate a mnemonics set in place X from another similar one set in place Y.

For example, if you set a mnemonics on Florence and another similar mnemonics on New York, you will be able to separate them without being able to confuse yourself. You will have no conflict with other images and associations.

Imagination

You will use your imagination in order to create the links between what you need to remember and what you have associated it with. Say, when you created images of a door with yellow Post-it notes, you were using your imagination. Thus, you want to allow your imagination to be creative and a bit crazy when you are trying to picture out things or keywords for association purposes.

Types of Mnemonics

Rhyme or Ode

"In 1492, Colombo sailed the ocean blue" — it is one of the most well-known rhymes to date. It happens to be among the many types of mnemonics that you can use to remember historical facts as well. Another usage of this technique comes when you need to recall rules of the English language, such as "I before E except after C."

Music

Writing lyrics or creating a little song can be helpful if you enjoy making music. Take a moment to think about how easy it is for you to memorize songs. You may even play parts of it in your head without relying on your radio or music players.

Acronyms

Acronyms are one of the most popular ways to create mnemonics. When you use an acronym, you are taking the first letter of each work and creating a saying with it.

For example, "Keep Educating Yourself" can be shortened as KEY, while TTYL is the acronym for "Talk To You Later." Chances are, you are using acronyms almost every day through direct messaging or texting.

Charts and Pyramids

Models are another type of mnemonics. The food pyramid, to be specific, teaches children and helps people remember what foods are more important than others. If you take a look at a food pyramid, you will see that whole grains and vegetables take up the biggest portion of it at the bottom, while sweets — the least important food group, which we can also eliminate from our diet — are at the top. As you look at each set, you will see their level of importance based on where they are placed within the pyramid.

Connections

Connections are another way to helps us remember things through mnemonics. For example, you may have been taught the word "long" when you are looking for the longitudinal line on the globe, which is the longest line that connects the North and South Poles. The reason why people remember the word is that it is the first syllable in the word "longitudinal".

Words and Expressions

Many people mistake words and expressions for acronyms, but they are different. When you are forming an acronym, you typically create a short word or abbreviation. However, when you are using a word or expression to help you remember things, you are using this type of mnemonic. For example, "Every Good Boy Does Fine Always" is often how music teachers teach the notes EGBDFA on the treble clef to the children. It is easier to remember the expression, after all, than a series of letters.

The order of operations in mathematics is another common example of this mnemonic. It goes like this: Parentheses, Exponents, Multiply, Divide, Add, and Subtract. Taking the first letter from each of these words creates PEMDAS. The thing is, the actual name for every symbol is almost impossible for people to remember easily. Therefore, the mnemonic that is commonly used is "Please Excuse My Dear Aunt Sally".

Acrostics

An acrostic is a poetic form that can be used as a mnemonic to facilitate memory retrieval. in fact it is a sentence in which the initial letters or syllables of each word are the initials of the concepts or words to be remembered.

So think of a sequence of letters to help you remember a set of facts in a particular order, like "Every Good Boy Does Fine Always", "Please Excuse My Dear Aunt Sally" but also "Mary's Violet Eyes Made John Stay Up Nights Pining" to remember the order of the planets.

8. Basic Memory Techniques

You might have trouble remembering names, numbers, faces, or what ingredients you need to get from the grocery store. Whatever it is, it seems to happen often, and you struggle to remember them every time. This can become frustrating for anyone. Fortunately, along with the techniques we have already discussed before, there are daily improvement strategies that you can also use to enhance your memory.

Write Down the Information

We have already mentioned in a previous chapter that you should write information down when you are building your observation skills. It will also help you build your memory in general.

In this day and age, it is hard not to sit down and type the information you need to remember. It is so much quicker to be able to open a Microsoft Word document or Google Doc and start typing up the information than keeping everything in your mind. It allows you to feel that, since you think about the information and spend your time typing it up, you are going to remember everything more easily.

The truth is, this is only helpful if you need to write

something quickly. It doesn't enhance your memory as good as writing the information by hand. Writing integrates multiple senses, touch, sight, and involves both short and long term memory at the same time. It stimulates the entire cerebral cortex and activates the faculties of attention and concentration.

The main reason why writing works better is that you are bringing brain cells that you no longer use to life when you start using your hand. These cells, which are known as the reticular activating system or RAS, tell your brain to focus more on the tasks you are doing.

Another reason is that, when you write, you are more likely to rephrase the information into your own words. Instead of typing information word-for-word, which people often do, you will think about what has been said and write it down in your own way. It will have the same meaning but with different words. Because you spend energy thinking about this, you will be more likely to remember the information.

Learn Like You're Going to Teach

Some people believe that the best way to learn something is to act like you're going to teach it. Whether you are trying to learn names or a series of numbers and memorize information for an exam, the more you believe that you will be teaching it, the more engaged you will become.

Another trick is to learn the information with the thought that you will need to teach a child. This will help you put the information into a simple form, which always makes any idea easier to understand and remember. As Einstein said, if you can't explain it simply, you don't understand it well enough.

Organize Your Mind

Many people feel like one of the best techniques to use, especially for beginners, is to organize your mind. When your thoughts are organized, after all, you will be able to remember any information better. This also brings up an important lifestyle choice, considering you may want to make sure that your area is clean and organized. The reason for this is that people often feel more relaxed in a tidy room. If you want to do it to your home, you want to do it to your mind as well.

Take a moment to think about how you feel when your desk, workbench, or kitchen counter is cluttered. It takes a lot more effort to focus on a task when there is mess everywhere, you see. Now, imagine how much easier it will be to perform this task if your work area is clean.

At this point, you may be wondering how you can work on making your mind more organized. After all, it isn't exactly like your desk at work where you can pick up an object and put it away. While this is mostly true, there are many tips and tricks that you can use to organize your mind.

Use a Written List

Once again, you can use a list in order to help your mind become more organized. In truth, people naturally feel more at ease when they have a list to depend on. For starters, it allows them to know exactly what they need to do. Besides, if you treat it like a checklist, you are able to cross out what you have done.

The point of this tip is that you only want to keep the information that matters. So, in a sense, you will discard anything that you no longer need to keep. This is why you need to use the written list method every now and then.

Be Consistent

Chances are, the items in your home have a certain place. For example, your coffee pot sits on your kitchen counter, your child's toy box is at the corner of their bedroom, and your silverware is in a certain drawer in the kitchen. This is the same thing you want to do with your mind. You want to make sure that everything has a certain place in it. For example, you will put the list of the 45 presidents into your mind palace in the form of a living room, while the list of everything you need to do before moving into your new house goes in the mind palace of your work office. As long as you need these lists, this is where they will be stored within your mind. Therefore, when you go through your list to make sure that you are officially ready for moving day, you can just imagine your work office and pull the information from there.

Be Aware of Information Overdose

We live in a world where technology seems to be in every aspect of our life all the time. It doesn't matter if we are using a laptop, tablet, or smartphone — many people are able to look up whatever they want whenever they want through their internet connection or mobile data plan. Because of this, our minds can become overloaded with information.

This can not only make us tired and stressed, but it can also push us forget about the important stuff that we need to remember as we deal with information overdose.

Being in this situation entails that your brain gets cluttered with a lot of unnecessary information. Aside from that, your mind is going to start soaking up everything like a sponge. In a sense, it will all seem like meaningless information to your database because it can no longer distinguish between what's important to remember and what's not.

Memory Hooks

Another basic way to help you upgrade your photographic memory is through *Memory Hooks*. This is a technique that is almost exactly as it sounds: you hook your memory, so you may not forget it easily. This follows the path that you are more likely to remember information that 'hooks' you.

Many people will use memory hooks on an emotional level. When people do this, they will anchor their memory to an emotion. This method works because our feelings can often serve as a trigger for certain memories. For example, if you remember almost being hit by a semi-truck on the road when you were younger, you might be cautious when you walk around similar or any vehicle. After all, your memory triggers an emotional response, which, in this case, is fear.

The stronger your emotion is that you attach with your memory, the more likely you are to remember what has happened in the past. If you had dinner with your sibling last week, for instance, you would probably remember having lunch with him or her and where you went to eat, but you might not remember anything more about it. You might forget what you talked about; if you did, you would have to think too hard about it and only get bits and pieces of information in the end.

Of course, you don't need to go through an event in order to use memory hooks with emotion to remember something. It doesn't matter what you want to call to mind, considering it can be a name, the address for your new house, or the definition of a word. All you need to do is associate an emotion with the information and match it up to a visual that is supposed to explain the associated feeling.

For example, if you want to remember your new house address, you can design the actual numbers as exclamation points because you are excited about your brand-new home. You can also make the visual a little more crazy by making the numbers jump as if they are excited about your new residence as well.

Three Important Pieces

In order to make memory hooks work well, you need to remember three important pieces of information.

1. The memory hook needs to be short and snappy. It is always harder to recall something that is a bit long and not interesting. Remember, you need to hook the information to your mind so that it knows to keep it in your memory bank.

2. The memory hook should be easy to remember. It won't help you if you try to associate the memory hook with an emotion that you don't often feel or doesn't fit well with the information. For example, if you want to remember the date and time for your surgery, you may not want to associate excitement with the event. However, this also depends on what type of surgery you are getting.

3. Only include the information you actually need within your memory hook. For instance, if you are trying to remember your new address but still live in the same town, you won't need to focus on remembering the town. Instead, remind yourself of the house number and street name.

Tips to Make Memory Hooks Interesting

How you will make memory hooks interesting will depend on your personality. Here are some tips to give you an idea of how you can create a hook for your memory.

1. Use puns to let people know what your business is. For example, if you are a dentist you may use a motto that sounds like "If you are not true to your teeth, they will become false for you."

2. Using humor is another great way to create an interesting hook.

3. Make a parody to make the hook interesting. You can produce one by taking a song and changing some of its lyrics so that they tie into what you want to remember.

4. Don't be afraid to mix and match or find your own way to make your memory hook extremely interesting to you.

Chunking Method

You can use the checking method for almost any long list of information. When you use this technique, you basically chunk or put together pieces of information. For example, if you have 10 numbers to remember, you can pair them up in order, which will mean that you only have to think of five numbers, which are similar to what your memory can hold when it comes to this information. For instance, if you have a list that consists of 8, 5, 3, 2, 1, 7, 6, 9, 4, and 7, you can pair the numbers as 85, 32, 17, 69, and 47. Take a moment to look carefully at this example and try to memorize the individual and combined numbers separately. You will quickly find that when the numbers are paired, they are a lot easier to memorize than the single digits. This also means that they are easier to encode and store in your brain, at least for a period of time.

Linking Technique

When you need to remember a list of names, you will often use the linking technique. It usually takes place when you need to link adjacent details on the list. You might recall taking a test with two columns in elementary school. The first column would contain a list of words, while the second one had the definition of some of the words in the first column. You would then have to connect the right word to its corresponding definition with a line. This method is similar to what you have to do when using the linking technique.

Three parts comprise the linking technique, which includes *creating* and *recalling* a list and then *practicing* how to do it repeatedly. Even when you are comfortable with the said method, try to take time to practice remembering one of your lists at least once a week. Otherwise, the list and the linking technique will start to decay and leave your mind.

The thing is, when you create any list, you want to make sure that each image or word links to the next one. For example, if you wish to write your grocery list, you start by getting your cart. You may then imagine the object that rests on the seat of your cart, such as a baby-shaped pineapple, assuming this is the first item on your list. In case the second object is a bunch of apples, you may imagine the pineapple as apples growing at the top. You will continue to link your list like this until you have reached the last item. It is important to recall everything in the same order to avoid forgetting anything on the list.

The next trick is to automatically remember the next item on your grocery list after picking up the first one. Because of this, it won't take a lot of energy to call to mind your entire list.

You need to take note of the fact that, when you practice the linking method, you don't have to feel the need to constantly practice remembering the same list. What you want to do is create a new list using this technique. For instance, if you go grocery shopping once a week, you may turn this memory exercise into one that's specifically for this activity.

This ensures that you will use this technique at least once a week. However, you can also use it throughout the week for other lists.

SEE Principle

The SEE principle is a memory technique that people often use to build their photographic memory from the start. SEE is an acronym, which stands for the three pieces of this principle: **S**enses, **E**xaggeration, and **E**nergize.

S Is for Senses

This principle states that the more you use your senses to encode information, the more you will be able to transfer the data from short-term into long-term memory.

E Is for Exaggeration

The second principle states that you want to be as creative, funny, and interesting as you can be when making your images, keywords, charts, graphs, or anything that you use to recall any information more quickly. Think of it this way: you are driving along the highway and notice a line of semi-trucks on the other side of it. You realize that one cab is all white, the truck next to it is white with a purple line, the third cab is pink, and the fourth one is completely white. You are going to remember all the pink-colored vehicles and the purple-lined white cabs more than the plain white ones because they are more interesting visually than the others. You would have remembered even more a vehicle with strange, funny and unusual drawings.

E Is for Energize

The last part of the SEE principle says that you want to make sure that the information you are trying to remember, along with how you want to do it, is energizing. For example, would you rather see a slideshow of the life of Prince or a movie about his life? You would most likely pick the movie over the slideshow because movies bring energy. There is movement in the latter, and you can latch onto the energy that you see the actors give throughout the movie. Films are better remembered because there is more involvement, more emotion and more excitement than other images. Create energizing images that you will hardly forget.

Memorization Tips

We all have things that we need to remember from time to time. While some of us find memorization easy, most of us tend to struggle with the process. If you are someone who feels too challenged when it comes to memorizing stuff, but also think it's not extremely tricky, know that you can use other extra tips. Here are some of the best ways to memorize information.

Prepare for Your Memorization Study Time

We all have different studying techniques. It is important that you take the time to get to know what you need to do to be able to study better. This will drastically allow you to improve your memorization skills. For example, you may find that you have to be quiet in order to remember your lessons more. If this is the case, then you need to look for an environment that does not give you a lot of distractions. Or considering you also notice that you need to have music in the background since the tunes help you to focus better, then make sure that you have the best music in place to boost your memorization skills.

Some people believe that it is important for them to prepare through a series of steps. For example, you may have to clear your mind of everything that you have learned that day. Therefore, you need to take time to watch a good film, have a cup of tea, read, or just relax. You may even find that you perform better when you meditate. If you need to play around with your preparations before you start to memorize, then you should do that accordingly to your schedule. Nevertheless, there is always time to change some of the steps as you continue to learn more about your preparation time.

Record and Write Down Information

Because writing down information is discussed elsewhere, I won't spend time on this. However, it is important to include it in this section as well. Say, if you think that it is better to record the lectures from your professors, then make sure to do this. However, you will also need to take the time to listen to the recording and write down any important information to be able to memorize what you need to know.

After all, not only are you hearing it, but you are also taking the time to get your brain cells active as you start to write some things down. Active brain cells always help you remember more information, too. Remember to prefer mind maps to standard notes. Mind maps are the most powerful tool you can use.

Write the Information Down Again

People don't realize how important writing down information is. In fact, many people state that one of the best ways to truly memorize information is to write it down when you first hear it and then write it down when you are remembering the information. In other words, write down the information from memory. Don't listen to the recording or look at what you have previously written, though. Instead, take a blank sheet of paper and simply go from your memory. Then, you can compare that to your original writing.

If you find that you need to continue to memorize the information, then feel free to do so. However, if you seem to be doing well with memorization alone, you can take a step back in order to test yourself a bit more. For example, you may not touch that information for a couple of days. Once these days are over, though, you can try to write down the same information from memory again and then compare the two writings. If you see that you are still going strong, continue to test yourself by lengthening the time interval. If you see that you have already started to forget things, then you should increase the amount of time you spend on memorizing the information.

Teach the Information to Yourself

Of course, you can teach someone else what you are trying to learn, but this isn't always possible. In this case, it is important to get into the habit of teaching the information to yourself. As you do so, you will find that you are more engaged when you memorize the details because you have the mindset required to explain or teach it. This is why you have to make sure that you understand the information before even trying this technique.

This is popular because it makes you more focused and gives you something to look forward to, kind of like a goal when it comes to needing to memorize the information. If you are like most of the world, you will need motivation in order to follow through with memorizing because very few people like to do this activity. This method, though, may motivate you to do what must be done.

Don't Stop Listening to the Recordings

A final tip is to not stop listening to what you have recorded. Many people feel that once they have listened to a recording once and written down the important information from it, they can already set it aside. Worse, they may decide to delete it or record a new lecture over it. Both ideas are not advisable, considering taking time to continue to listen to the lectures is going to help your memory improve through its own technique. Repetita iuvant. Repeated things help.

9. Advanced Techniques

Before I start to discuss more advanced memory-enhancing techniques, you may feel like the methods discussed here or in the previous chapter are either basic or too advanced for you. It is always easier to start with some of the simpler methods — the ones you feel are easier — and work your way up from there. This is something that no one can directly tell you as it depends on your personality and where your memory already sits.

Another factor to remember is that every technique is going to seem hard for you at first. However, once you manage to try it successfully a couple of times, you will soon be able to get the hang of it.

The Car Method

The car method is similar to using a room in your home as a memory palace. One of the biggest reasons why it is considered as one of the more advanced techniques is that some people don't know the parts of a car. Furthermore, they can become confused as they don't see the car in the same way as a room in their house. These individuals may feel that going from the trunk to the front of the vehicle is a bit more confusing than going around any room. As stated before, though, the level of confusion depends on your personality and interests.

At the same time, the car method is highly useful because many people have an auto that they can use to observe instead of just visualize. Similar to using a room in your home, you will want to make sure that you know your car well, as well as everything in it, before you start using this technique. For example, you should familiarize yourself with the storage compartments because these are often the places that people may this method for. Cars, especially the newer models, can have a dozen of storage units all over them. Not only are they on the side of the doors, between the seats, and on the back of the seats, but they can also be hidden in the trunk of a car.

Of course, if you don't have a car, you can use any type of vehicle that you know well, such as an airplane, bus, or semi-truck.

Another example that you may want to look at is a list of animals in a reserve, which looks after injured and abandoned animals before returning them to their natural habitat. You can use this information to make sure that you and your family will be able to see them all without having to check the map all the time. Plus, knowing the list by heart allows you to create a game with your children in which you ask them to find or name the animals there. Hence, you may use the car method to memorize the following animals: penguin, llama, tiger, bear, eagle, buffalo, wolf, duck, and otter.

You know that the penguin is the first animal that your children will see. Therefore, you want to imagine the penguin at the front of your car, considering you wish to remember this list from the front to the back. You may imagine a penguin sliding on the hood of your car. From there, you want to connect this image to a llama, which may be driving it. The tiger is perhaps sitting in the passenger's seat, while the bear is trying to fit in the pocket at the back of the driver's seat. Feel free to continue to use this list with the same method in order to memorize the rest of the animals in the reserve by the order in which you will see them.

The Peg System

The *Peg System* is another common technique that seems more advanced for some people. When you think of the pegging method, you may think of clothing pegs. In truth, they are a bit similar to each other. This technique use visual imagery to provide a 'hook' or 'peg' from which to hang your memories.

This system works by creating mental associations between two concrete objects in a one-to-one fashion that will later be applied to to-be-remembered information. This method works by pre-memorizing a list of words that are easy to associate with the numbers they represent. Those objects form the "pegs" of the system. Typically this involves linking nouns to numbers and it is common practice to choose a noun that rhymes with the number it is associated with.

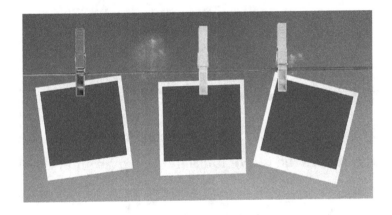

A complaint about the Peg System is that it seems to be applicable only in trivial situations. However, the Peg System can be used to remember shopping lists, key points in speeches and many other specific lists to one's areas of interest.

With this method you will easily remember the numerical position of items in a list in sequence or out of sequence

Why Use the Pegging Method

The pegging method is known to be one of the commonly advanced techniques for several reasons.

1. There is a lot of flexibility between lists

When you are able to create flexibility with lists, you can reduce the risk of interference. For example, you can use ordered or alphabetical lists to associate with the pegging method. Of course, many people suggest that, when you start using this technique, you should pick a list that you are more comfortable with, such as an ordered one. After using the pegging method a few times and understanding how it works, then you can go with different types.

2. Some people don't memorize items well

If you find that you struggle with memorization, you may realize that this method may not be extremely helpful for you. The reason is that you need to keep order, which memorization does not always provide. Aside from that, it allows you to use whatever list that comes to mind.

3. You can directly recall the item

While the *Linking Technique* is ideal for remembering lists in sequence it does not provide an easy way to recall, for example, the 7th item in the list. You should start at the beginning of the list and mentally count forward through the associations until you reached the 7th item.

You can have 20 animals in a specified order that follows the reserve's map, for instance. If you wish to pick out the seventh animal, you will need to go through the entire list starting from the first one until you reach animal #7. Instead, with the Peg System, you can directly remember the article, for example: Seven = Chicken

There are several lists that you will memorize through images enough, and you won't always have to keep the order. For example, if you are trying to create a list with the animals of the reserve, you can eventually pick the animals by yourself without having to go through the whole list.

4. You can use the Peg System to hold more information

As mentioned above, the peg system offers a lot of flexibility. In truth, you can mix it with other techniques that you have learned. Use your favorite basic method or another advanced technique, along with the peg system, for example. By doing so, you can open the door to being able to encode, store, and retrieve more information than what you can do through one list at the time.

One of the common peg system lists is the alphabet system. If you use and mix it with the linking technique, you can remember over 200 items in a single list. While it may not seem possible now, you need to remember that you won't place all the items in your list at once. Like many lists or mind maps that have gotten larger than life, it is something that you can build over time.

Rhyming Peg Method

If you like rhyming, you will enjoy the rhyming peg method. The idea is that you need to create a list of words and then find other words that rhyme with them. For example, if you have a duck on the list, you can rhyme it with truck. Pig rhymes with dig, dog rhymes with jog, cat rhymes with bat, etc.

But usually a list of numbers is created, and we match words in rhyme, for example:

0 = hero

1 = bun

2 = shoe

3 = tree

4 = door

5 = hive

6 = sticks

7 = heaven

8 = gate

9 = wine

10 = pen

The fun part about the rhyming peg method is that you will be able to improve your creativity with it. Say, you can give the rhyme a beat and create a silly song or make a story in which you start a sentence with a specific word and then end it with a rhyming word. The more creative and fun you become with this information, the easier it will be for you to recall the information when you need it.

Alphabet Peg Method

Within the alphabet peg method are two types of lists that you can create: *sound alike alphas* and *concrete alphas*. Of course, you can get creative and establish your own as you become comfortable with the process, but now let's see these two types.

1. Sound alike alphas

The sound alike alphas list is no different from the rhyming peg method, but you will have to find a letter that sounds like the word. For example, B sounds like a bee. Therefore, you may imagine a bee that's shaped like the letter b.

2. Concrete alphas

When you create a list of concrete alphas, you will go

through the alphabet and find a word that starts with the corresponding letter. It is not necessary to make the words rhyme; you don't have to worry about the sound or give the words silly shapes or images either. The list that you create will be useful when you are trying to memorize certain information. For instance, you may put together an alphabetical list in which A stands of Apple, B stands for Bass, C stands for Cord, D stands for Drum, and so on.

Shape Peg Method

This method is similar to the other methods, although its main distinction is that it uses shapes. Basically, you will turn the information that you want to remember into a certain shape. The figure may correspond with the word or is perhaps the first shape that comes to mind when you think of it.

Spaced Repetition

Many people, especially beginners, feel the need to repeat the information to themselves in order to remember it. Unfortunately, this will only work for a short period. You need to keep in mind that that, unless you use a method, you are emotionally attached to the information. It is also possible that your mind believes that it is important for you to remember something that you will most likely forget within a couple of months or so. It doesn't mean there is something wrong with your memory. It is normal for people to start forgetting information that they don't use or recall over time. The primary reason why this happens is that your brain is making room for more important data that you will need to remember in the future.

Therefore, a lot of individuals, especially the ones who often practice memory-enhancing techniques, state that they often focus on recalling the information that they want to keep at least every couple of weeks. This is a great method that many competitors for memory contests tend to use. After the competition, they do not train their brain for a few months. Then, a couple of months or so before the contest, they will start to train their brain again. Once the process begins, they will not only use a variety of techniques — such as timing themselves — but they will also practice with different lists weekly, if not more. This helps them in many ways.

For one, it allows memory game players to improve their speed, which is a big factor when it comes to contests. Secondly, the practice helps them retain old and build new information in their memory database through a different method. For instance, they may recall a list from last week and then focus on learning a new list the next.

Of course, you can try the spaced-out repetition for six months and not touch the list until you have to. The gap will mostly depend on your ability to recall the list; that's why the training may also take longer than that. Many people state that if you have lists that you always want to remember, you will need to follow the spaced-out repetition method with each of them. This ensures that you will be able to keep every information fresh in your mind. In my book

"*Accelerated Learning*" I reveal my personal study system that I use to memorize information forever thanks to Spaced Repetition.

Memorizing a Deck of Cards

Another great technique that many beginners use to boost their photographic memory is memorizing a deck of cards. If you are just learning how to build your memory, you may feel like this is an impossible task because there are exactly 52 cards within a deck. However, almost every person who has stepped into the advanced photographic memory training has had to practice with a deck of cards. After all, cards are easy to get ahold of. In fact, you may already have a deck of cards in your home. Apart from that, they are already designed, have numbers, and are color-coded; that's why they can make the learning process a bit easier when you are trying to enhance your memory.

There are a few basic things that you need when you

are going to memorize a deck of cards, aside from making sure that you have a complete a deck of cards. You must also have a list of 52 celebrities — ones that you like and don't really care for — and knowledge of creating a memory palace.

First, you should understand that, when you are learning a deck of cards, you have to use a technique similar to this. The reason is that without a proper method in place, it will take you at least half an hour to remember half the deck of cards. On top of this, because you have not associated the cards to anything that feels interesting to you, the information will more than likely be forgotten over time. In fact, you can forget everything that you have memorized within a couple of weeks.

Create a Memory Palace

Most people will think that they need to memorize the cards based on the numbers and designs. While you can do this using another memory technique, this specific method doesn't focus on such things.

Instead, you have to concentrate on the list of 52 celebrities that you have written down.

In order to make memorizing cards as easy as possible, you can categorize your list of celebrities with the symbols that are already on the cards. For example, diamonds can be used for the wealthiest celebrities that you have on your list. The hearts may match the celebrities that you love, the spades are for the ones you don't really like, and the clubs for the celebrities who seem to party too much.

Then, you will want to pair up your celebrities with even or odd numbers. From my experience, it is always effortless to show that the men are the odd numbers, while the women are the even numbers, or vice versa.

You can then use the members of the royal family for the king and queen in the deck. For example, Queen Elizabeth will be the Queen and Prince Philip will be the King. For the joker, you may use Jack Nicholson or Heath Ledger, considering both played the Joker in the Batman movies.

From there, you can match celebrities up with numbers. For instance, you may feel that the 10s should be the most powerful celebrities on your list. For the 9s, you may decide that they should be your favorite celebrities, 8s may be musicians, and 7s may be athletes. Everything depends on how you have listed their name. This is the best way for you to memorize your deck of cards.

Memorizing and Recalling

Once you have organized your list and corresponded them with your cards, you will then start memorizing your cards. In reality, you can use a memory palace or even a mind map in order to do this.

It is important to realize that you do not have to memorize all 52 cards at once. In fact, you can create a memory plan that will build up to memorizing all the cards. You may start with five cards every day, and that's okay. However, you also want to recall the cards that you have memorized before. So, on your first day, you will focus on the first five cards. On the second day, you will recall the first five cards and then memorize the next five cards. You will do this until you reach the last seven cards.

The Military Method

While the steps associated with this method are simple, the debates on whether the military technique works or not, are more popular than the method itself. Those who have never tried this technique should not talk. Some military units have been using this technique for almost a century to develop their photographic memory.

You have to start by being in a dark room with a lamp next to you. You also have to have a sheet of white paper with a cutout just large enough to fit a paragraph of text. Then, get a sheet and cut a rectangular hole out of it about the size of a standard book paragraph and then place it on a page of a book.

Adjust your distance from the book so that your gaze instantly focuses on the words when you open your eyes. Stay in the dark for a while to accustom your eyes to the dark and then turn on the light for a fraction of a second and turn it off again. You will have a visual imprint in the eyes of the text that was in front of you.

When this imprint fades, turn the light back on for a fraction of a second, and fix the text again. So, in a nutshell, you'll be sitting in a dark room, and you'll turn on and turn off the lights to memorize and see in your mind the imprints of the text you're reading.

Keep doing this until you can read the text word by word. When you see the imprint in the darkness, you are not seeing the text in the dark, rather your brain remembers a virtual imprint of information and this is the idea behind the memory of the text.

Would you like it if you could develop the ability to quickly look at a piece of text and be able to see the imprint in your mind? The thing is, you will need to do this for at least 15 to 20 minutes every day for 30 days. This enhance your ability to glance at an image or passage of text and memorize it instantaneously.

10. How to Remember...

It doesn't matter who you are, you will always struggle to remember something, whether it is a person's name, a place, what your children's favorite meals are, or anything else. This is why it is important to strengthen your photographic memory with the use of the techniques that we have discussed previously. By now, you have probably tried some of them and may already have an idea of which ones you are comfortable doing and need to be practiced a bit more.

If you haven't taken the time to build your first memory palace yet, you should try to do that soon. While it is not essential for this chapter, the earlier you start building your photographic memory, the more you will be able to recall pieces of information that we are going to discuss here.

There are two main parts of this chapter. The first

one involves learning how to remember names. It has happened to all of us. We meet one of our significant other's family members at a family reunion. Then, a few months later, you recognize the person at the grocery store, but you cannot find their name in your memory bank. Of course, this is a bit embarrassing for you because they remember yours. When this happens, you will often dance around the idea of how to let them know that you don't remember their name. You act as if you do, but you never say their name or ask about it. Instead, you go home and ask your significant other what that person's name is. Of course, this also helps us remember their name a bit better. Don't worry, this is a human thing. While we may forget a name initially, when we run into the same individual and need to exchange pleasantries with them again, we are more likely to remember their name because we feel like we have made a mistake and don't want to commit that again.

The second part is remembering numbers. It seems like people used to remember numbers better before the creation of cell phones. Now, we tend to struggle a bit more with this activity because it is so much easier to add the digits in your contact list than memorize them. However, what is going to happen when you leave your phone in the car and don't have a piece of paper and a pen to write down the number of a person you just met in a store? Or, you're at the grocery store and forgot what your partner asked you to take and you have your mobile phone in the car. Of course, you can run back to the parking lot, but then what will you do with your cart that's full of groceries? You can give a stranger his/her number so that they can call it on your behalf, but do you even know his/her cell phone number? If you are like several other people out there who are not 100% sure about what their cell number is, it is obvious that you are practically doomed.

Remembering Names

Oh, the wonders of name tags! Have you ever had to

be in a large group of people and found that name tags were a big help when it comes to recalling the names of each person there? Do you remember starting your first day of school and not only going around the room to introduce yourself but also having your name on your desk and perhaps being given a name tag to stick on your shirt? Or you may you have learned about your child's new classmates by looking at their name tags. However, it doesn't mean that you will remember their names when you run into them again at your children's school play a couple of months later. You might be able to recall where you met and talked, that they were wearing a blue suit with matching blue shoes, but the name might have already escaped your memory.

You may also remember something about the person's character. For example, while they were sitting across the room, you were able to hear almost everything they said due to their loud voice.

All of these examples are ways to connect someone to their name. The first one is known as meeting place connection, while the second and third samples are called appearance and character connections, respectively.

Meeting Place Connection

When it comes to meeting people at a specific location, you can use this place in order to help you remember their names. This is a technique that you will use, sometimes through your subconscious mind, to create an automatic association. Nevertheless, it is no indication that the subconscious will become conscious when you need it. All of this will take place automatically within your mind. However, you can also associate another place with certain individuals by yourself.

When you are looking at a meeting place connection through your conscious mind, you are trying to find a way to associate the person's name and face with the location you are at. For example, you are at the park, and your daughter starts playing with another girl around her age. You go up to the other little girl's mother and introduce yourself. You then find out that the mother's name is Clarissa while her daughter is Alessandra. As you talk to the mom, you are trying to come up with a way to remember their names, as well as where you have met. You think of how the name Clarissa sounds like a beautiful word and then connect it to the park because you believe it is a beautiful place.

A couple of months later, you are going for a walk with your daughter who starts waving at a couple of people walking towards you. You recognize their faces, but you don't remember the names. You then start to think of where you have seen them before and recall that it's at the park. This is when the word 'beautiful' comes to your mind, and you remember the mom's name is Clarissa. From there, you are able to remember that the daughter's name is Alessandra. By the time you meet the two on the sidewalk, you already know their names again.

This situation can also happen subconsciously. For example, through your unconscious mind, you may be able to simply place the faces within the park and then remember the names. This means that no thought on your part went into associating the names to the park; instead, it all happened within your mind as you were talking to Alessandra's mother, Clarissa.

Appearance Connection

Just like with the meeting place connection, you can associate names and appearance either subconsciously or consciously. When you use appearance connection, you will be connecting a part of the person's physical appearance that you find interesting to their name.

When people use appearance connection, they are careful to observe all the person's physical characteristics. While you can use something like what the person's wearing, especially if it really stands out, it is more common to use physical traits, such as hair color, eyes, smile, etc.

Say, you are heading to your local historical society and museum because you need to talk to one of the employees about donating old documents that your great, great grandparents brought over when they immigrated from Norway to the United States. When you walk into the museum, you meet a girl who's sitting at the admissions desk. The first thing you notice about her is that she has purple hair. As you start telling her your reason for visiting the place, you find out that her name is Valentina and that she is the person you need to bring the documents to. You tell her that you will bring them to the museum in a few months when you come back from your trip. She tells you that, when you bring them in, just tell whoever is sitting at the admission's desk that you need to see her and that you won't have to pay the admission fee if you don't want to tour the place. Then, you thank her and leave.

Upon returning to the museum after a few months, you realize that you don't remember the employee's name. However, you know that someone will be able to tell you who to talk to. When you are walking into the museum and see a man sitting at the admission's desk, therefore, you remember that a woman with the purple hair used to sit there and that her name was Valentina.

Appearance connection can also work if you meet someone in a different place. For example, you have come back from your trip but haven't made it to the historical society and museum yet. However, as you are grocery shopping, you notice someone whose face looks familiar. She smiles at you and then you notice her purple hair. You then remember that it is Valentina from the museum.

Character Connection

Character connection works like appearance connection; however, instead of remembering someone's name because of their physical features, you can recall something special about their character. Like the other forms of connection, it can happen subconsciously or consciously.

Let's say that you meet someone by the name of Roger Nelson while you are in the grocery store. You started talking to him as you were waiting in line for the cashier, who was trying to get the cash register fixed. Neither you nor Roger was in a hurry, and you didn't mind waiting at all, so you let other people between you go ahead at the other cash registers that were open and working.

As you started talking to Roger, you learned that he was teaching psychology at the local university. You also found out that he has three children who go to the same school as your kids. In fact, his son is only a grade above your daughter. As you continue talking to him, you learn that Roger is about to take a trip to Italy. You have been to Italy, so you start telling him what places he should see. As the register starts working again and he begins to check out, you learn that he also just moved from London, England, which is why he has a thick accent.

A few months later, you are at your daughter's school play when you see a man with a familiar face. He smiles and starts talking to you. This is when you recognize his accent. You then remember that he's supposed to go to Italy, which then makes you realize that this guy's name is Roger. As all the information that you have previously learned about him comes back, you ask him about his trip, how he is enjoying your town, and if he misses London.

In this example, you will see that you don't have to simply associate a name with one characteristic. The truth is that you can also do it with parts of a whole conversation. Only, how you associate the name through a character connection will depend on what you may or may not find interesting about the person.

Remembering Numbers

When it comes to numbers, the average person can remember between five to nine numbers. While most people don't tend to focus on improving their memory with numbers, it is just as important as names. This is because digits are everywhere in our lives. Not only are there phone numbers, but there are also house, account, and bill numbers. In fact, if we want to pay for something online, you will need to supply the figures on your debit or credit card. How often have you been asked for your credit card number but can give it right away because you don't have it with you? Instead, you have to go to your room to get the card from your wallet.

Or you are on the phone with an operator to activate a service and you need to provide personal data, and even in this case, not remembering them, you have to go and get them in your room. If you have been in the same situation, you know how annoying it is not only for you but for the person on the other end of the line as well. Everyone has their own busy lives, so the faster you are able to give the caller your personal data, the faster you can focus on something else.

As stated before, you don't want to focus on repeating numbers continuously for a period of time as they will more than likely end up in your short-term memory. While this will be fine if you decide to write down the number, it can often make us feel like we have repeated the number enough to remember. Despite that, when the time comes and you have to retrieve it, you fail to recall parts or all of the number. Therefore, you need to try other techniques that will allow you to transfer the digits from your short-term memory into your long-term memory. It is something that you have to practice often so that the information in your mind will not start to decay within a few months.

From the get-go, I will let you know that you can use the rhyming peg method in order to remember numbers. Because we have already discussed this technique, I will not explain it again. However, I felt like it was important to mention it here again because people commonly used the method when they want to recall digits.

Here are a few other practices that you may try.

The Journey Technique

One of the techniques for remembering a long series of numbers, such as a credit card number or account number, is *The Journey Method.* This is similar to creating a memory palace. However, instead of using a room, you are more likely to take yourself on a journey. For example, if you drive for half an hour to work five days a week, you may say that this is your journey. You will start by observing the path thoroughly in the morning, so you will become mindful of all the landmarks on your way. From there, you will be able to associate a number with each landmark. This technique combines the narrative flow of the Link Method and the structure and order of the Peg Systems into one very powerful system.

This technique is helpful when you often take the route because you can remember the associations well. On top of this, you will start to become more aware of your surroundings as you are driving to and from work.

Number Shape Method

There are a couple of ways to use the *Number Shape Method*. While the main factor is that you want to associate a number with a letter, you can decide on what shape the numbers will take the form of. For example, because the number 5 looks like an S, many people tend to link the two with each other. However, when it comes to the number 1, you can choose between T and D. Of course, you can also decide to associate the L with 1 as well. With so many possible matches, though, you may want to write down the list.

Because there are limited shapes, many people like to associate the numbers with the shapes of letters. However, you can also choose to create a list of shapes and associate them with numbers. You usually have to match the first 9 numbers plus the 0 (zero) with shapes because you can simply double the shapes if you have a double number. If 0 is a circle and 4 is a star, for instance, in order to say 40, you can put the star and circle together.

Other people like to associate the numbers with the letters because there are 26 letters and 9 single-digit numbers. This means that you can link more than one letter to a number. This often helps people to remember keywords or phrases. They will also use this system to recall parts of a story that they have heard in the past. For example, you can make the word GOOD by saying 6 looks like a G, 0 looks like an O, and 1 looks like a D.

11. Continue to Build Your Memory

Photographic memory is not a gift you were born with. You were born with your memory database, but you need to use mnemonic techniques to enhance it. Furthermore, photographic memory is similar to using a muscle. If you don't continue to use it, it may become immovable sooner than later.

Therefore, it is important to make sure that you continue to build your memory through different methods. This is often the reason why people start with basic strategies and then move on to more advanced ones. They are slowly increasing their photographic memory instead of forcing it to fade away as quickly as possible.

Tips to Help You Become Successful

There are a lot of factors that go into helping you improve your photographic memory. Not only do you need to use methods, but you also have to know some information on how to become successful as you use them. This is what these tips are for. They are here for your benefit, so you can reach your full potential as you enhance your photographic memory.

Stay Focused

One of the biggest struggles for people who are working to improve their memory is that they cannot stay focused. They may let their mind roam while trying to work on techniques or recalling information. Worse, they may start to become bored with a certain method.

Sometimes, you need to realize that if you are getting

bored with the technique, you shouldn't be focusing on it. Your concentration may be suffering because you are uninterested in that technique. This is the nicest part about having so many methods to tap into, indeed, we can pick the most interesting ones and choose what works for us.

Another reason why you are probably struggling to remain focused is that you have been working on or practicing the same technique for too long. While it is good to train yourself, you want to make sure that you are not doing it too often. In fact, some people suggest that you should take time every day to focus on improving your memory, but you don't want to overdo it. If you focus too much on a single method, you are going to start to feel tired and overwhelmed and lose interest in it. This can later make you feel like you shouldn't try to improve your memory at all. To avoid this problem, you should take everything in stride and take a break whenever you have to do so.

The biggest problem that you may have with a break, though, typically comes if you are in the middle of creating a memory palace. Most people will tell you not to break away when you are doing it because you will most likely need to start over. Depending on how strong your memory is, you may still be able to take a break in the middle and then start again once you have more energy to finish your memory palace. However, if you struggle with creating one from the start, you have no choice but to complete it without a break.

In reality, the decision depends on what you want to do. One factor to think of is if you will be able to remember your mind palace creation when you are struggling to remain focused.

If you think that you will have a hard time keeping it in mind when you go back to recall the information, then stop focusing on it and let it go at once. In case you don't want to give up, you can always take the time to write the information that you have gone through. It can help you remember things when you back to finish your mind palace.

Take Time Every Day

The only way you will really improve your photographic memory is if you take the time every day to work on your memory. Remember, you want to focus on building your memory slowly since this will allow you to recall the information that you have previously stored in your mind and help you feel more at ease when the memory-building process starts.

At the same time, the more you try to force yourself to learn at a fast pace, the less likely you will be able to recall anything.

Think about how you studied for your exams at school one time. If you crammed under pressure, you probably didn't remember your lessons well, even if you tried to memorize some of them. The same thing is true when you try to cram a lot of memorization techniques in a short amount of time instead of learning them slowly yet steadily.

Don't Allow Yourself to Procrastinate

One of the biggest keys to making sure that you can improve your photographic memory through these techniques is to prevent yourself from procrastinating. You want to be efficient, especially if you are using a few of these them to memorize any information that will appear on your exam. After all, when you procrastinate, you will find yourself needing to learn stuff quickly and within a short amount of time. You will then feel like you are forcing yourself to cram everything in your brain, which, as stated above, is not what you should be doing.

Furthermore, by procrastinating, you will sense that all of your work is piling up at once. While you used to have enough time to learn everything, due to procrastination, you are now feeling stressed. As you probably remember, stress will negatively affect your memory, especially if it is chronic. There are some people who can function well during exams when they are only dealing with acute stress.

Unfortunately, many people live such busy lives and have so much going on that they are naturally stressed. Therefore, when we add something else to the mix, we will only become more stressed than usual.

Find Techniques to Concentrate Better

While we have already talked about the need to stay focused, it is now time to talk about things that will allow you to make it happen. However, finding techniques to make sure you are able to concentrate can happen whether you struggle with your focus or not. For example, many people can focus more if they can hear background noise. If that's the case, you want to play music while you are working since this will motivate you to complete a task. At the same time, others feel like they can't do this because the sounds can interfere with their ability to remember things. Hence, in this case, music may not be the best concentration tool for you. You can then find another technique to maintain your focus, such as walking around or jotting information down, meditate or stay in a place alone.

Always Remain in Control

There are times when we feel like we are losing control. When this happens, we can start to feel chaos inside our heads. This isn't good when you are trying to learn techniques to improve your photographic memory, though. If your mind is not structured and organized, you may not be able to remember all the information you see. It will make you more frustrated as you try to memorize stuff using different techniques, which can then result to other problems. Therefore, the more you feel like you are in control, the more you will be able to become successful at remembering things.

Practice Self-Discipline

Many people forget the distinction between discipline and self-discipline, which is often the reason why they don't remember to become self-disciplined when it comes to their lifestyle. However, this is one of the most important tips that you will find within this chapter.

When you try to teach self-discipline to yourself, you are trying to get yourself to behave in a certain way. For example, if you want to take time to practice your photographic memorization techniques every day, you need to tell yourself that you have to do this. Even though you are tired or uninterested in practicing your recollection skills for 5 or 10 minutes, you will do it anyway because you have already conditioned yourself to do so.

When it comes to self-discipline, there are a lot of important steps that you can follow to master it. For one, you can look at this list as a series of steps that you have to accomplish or see them as tips that can guide you towards your goal of becoming a self-disciplined individual. Whatever you decide to do, it is important for you to know that once you start to become self-disciplined, you will notice a change throughout your day. After all, self-discipline will not just focus on your memorization techniques but also on other factors within your life, such as exercising, eating right, and getting up when you set your alarm.

1. Make sure that you have a goal or vision in mind

You want to know exactly what you are working towards, so you have to make sure that you are aware of the self-disciplining techniques that may help you advance your memory. You might be doing this for your daily use, to help you decrease your chances of a disease, or because you want to join a memory competition. Whatever your goal is, you need to have something to work towards; otherwise, your effort may fall to the wayside.

2. Try to develop your self-discipline with a friend or family member

Chances are, you know another person who needs to

improve their self-discipline. You are more likely to continue to work towards something if you have someone who's doing the same thing beside you. You are also less likely to become bored if you can turn this into some type of competition with a loved one. Nevertheless, if there's no one you can do it with, you can make daily goals that you have to accomplish before moving on to the next one.

3. Be 100% committed to developing your self-discipline

It is typical for someone to come up with an idea, think that it is great, want to accomplish it, but then realize that this idea isn't really something you truly care about. Because of this, you find yourself uncommitted to the task that you have started. Sometimes, you may try to continue working on it, but once it starts to feel forced, you may realize that you don't want to work on it at all. Other times, you will find yourself taking a break and then forgetting what you have already done, so you have to start over. Nonetheless, because you are not committed, you are not sure about what you want to do.

Before you start to put any work towards developing your self-discipline or improving your photographic memory, you need to guarantee that you are completely committed to the task. By now, you have read most of this book and probably already decided where your commitment stands, so stick to it.

4. Remember that the more you make yourself accountable to accomplishing your goals, the more you will want to work toward them

Many people don't think about making themselves accountable for their actions. However, if you do so, especially when your focus is on developing your self-discipline, you are more likely to accomplish the tasks that you have set for yourself. Now you have all the tools required to become responsible. All you need to do is use them. Making yourself accountable for your actions is a great way to showcase this.

You can also hold yourself accountable by establishing a rewards system. For example, if you complete the task you set for yourself that day, you can watch a good movie. In case you fail to do, you need to restrain yourself from even logging into the platform.

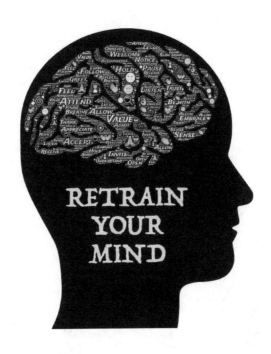

12. Practice Makes Perfect

You can think of this chapter as a bonus one to help you get started on a couple of techniques. I will walk you through a couple of them that we haven't officially discussed yet. It is my hope that, through this chapter, you will be able to start improving your photographic memory at your own pace.

Exercise #1: Remember Names

Read the following story and use the three connection techniques — meeting place, character, and appearance — to remember the name of the presenter.

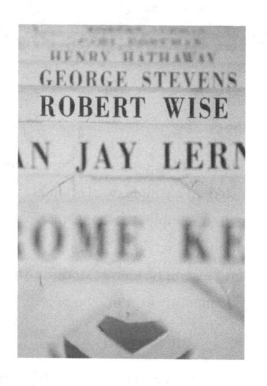

Donnie was running late when he got to the building for the presentation. He was there on behalf of his supervisor. While Donnie had never met the presenter, his supervisor was good friends with him. Because Donnie was running late, he didn't care to pick up an information packet by the door, which could have shown him the name of the presenter. He got into the room and quietly sat down as the presentation was already starting. When it was over, Donnie took his turn to meet the presenter. The first thing he noticed, however, was that the man was dressed in a brown suit and blue socks. Donnie also saw that the presenter had a lip ring and a big wedding ring on his finger.

"You must be Donnie," the presenter said with a thick New York accent. "I'm Fred Matthews. It is a pleasure to meet you." Donnie smiled and spoke briefly to Fred before walking away and going back to work.

Exercise #2: Memory Palace

For this exercise, you will focus on creating a memory palace. Of course, if you have already created one and are not comfortable with the idea, you don't need to do this right away. However, you should still try to do this exercise when you are ready to create your next memory palace.

At this point, you are going to focus on a room in your home. You will also make a list of the techniques that you can use to improve your photographic memory. You will be able to associate keywords with an item in your mind palace. For example, if you want to become more patient because you know that you will struggle with the slow and steady process, your keyword can simply be 'patient.' If you need to limit stress, you may use 'stress' as the keyword.

Before you start, write down your information. This will help you make sure that you are using a certain order, maybe from the most to least important. You should also write the keywords so that you don't have to think about all of this as you come to the next item in your memory palace.

Bonus Technique: The Emotional-Based Approach

By now, you know that emotions are a large part of

being able to remember information. After all, our brain is more likely to store data when they are anchored to feelings. This doesn't mean, however, that you have to attach emotions to every information that you wish to keep in your memory bank. There is one technique that shows you how important emotions are when it comes to your memory.

In order to attach an emotion to some details, you have to actually feel it. When you are thinking about a situation, for instance, you should experience it. At the same time, you need to remember that your brain doesn't multitask as well as people think it does. It is a lot better for your memory if you focus on one piece of information at a time. This way, you will be able to make the connection better than when you try to feel the emotion.

Now, I am going to give you a story that is full of emotion. As you read it, I want you to get in tune with your own feelings. Imagine how you would feel if you were the girl in the story. You should also picture out what she looks like, what her facial expressions are, and what her mannerisms may be, among other things. Think of it like a movie in your mind as this idea will help you get in touch with your emotions easier.

It had been over a decade since Alessandra stood at the doorway of her grandparent's farmhouse. She allowed her mind to wander back to the time when she was 15 years old and putting her band instrument away. While Alessandra was placing her clarinet in the shelf, she heard the school secretary say over the intercom, "Mr. Cardinale, could you please send Alessandra to the office?"

Alessandra waved to her teacher as she walked toward the office. The whole time, she wondered what she had done. Alessandra was a good kid and had almost never gotten in any trouble. As she turned to the corner of the hallway, she saw her mother standing right outside of the principal's office. She was about to ask what happened when her mother told her with tears in her eyes, "You need to come home, your grandpa has had a heart attack and is in the hospital."

Alessandra stood there for a few seconds, trying to come up with words. The only thing she could think of saying was, "Grandpa?"

Her mother nodded as Alessandra kept repeating that word in her head. She slowly walked back to her locker to grab her backpack the square and the compass. Alessandra kept telling herself that it had been her grandma who was sick all these years. How could her grandpa, who seemed healthy, have a heart attack? Besides, he was still young. He was only 68 years old.

The following week, Alessandra's grandpa passed away. Now, 12 years later, Alessandra came back to the house. She hadn't been there since a few months after her grandpa died and her family came to pick up the furniture for an auction. She ran her fingers over a crack in an old wooden cabinet. She then took a couple more steps into the house. The first thing she could recall was how her grandpa used to play guitar in his upstairs bedroom, but it would be heard all around the house. Alessandra smiled as she remembered running up the solid steps towards his bedroom and sitting next to him on the bed as he would start singing a silly song to her.

Alessandra then looked at where the dining table stood in the kitchen. She remembered how it always held a big meal on Sundays. Everyone would come back then as there would be bruschetta, pasta, chicken, dressing, roast potatoes, celery, spicy sauces. She took a deep breath as she could almost taste the food.

Alessandra continued to walk through the house. Sometimes, she would stop and think about some memories of her childhood. Other times, she would look at how much the place had changed, especially all the empty alcohol bottles from when people had partied there. She started picking them up until she noticed the bedroom in the corner. Ever since Alessandra was little, she never liked the closet in that bedroom. While she wanted to walk in for just a quick minute, she also didn't want to see that closet. Alessandra never understood why that closet made her feel uneasy. Either way, she wanted to *focus more on picking up all the empty bottles because they did not belong in her grandpa's house.*

Nevertheless, as she picked up one bottle, Alessandra came to the realization that it really didn't matter anymore. While this place still belonged to her mother, it was also a party house, whether she liked it or not. No matter how many beer bottles she picked up, she would continue to find more when she came back to visit.

As Alessandra walked back to her car, she took one last look at the house and the yard. She saw the old swing set and smiled. "I had a wonderful childhood," she told herself before driving off.

Conclusion

There is a big debate in the psychological field about whether photographic memory exists or not. Some people state that it doesn't because we manipulate our mind into remembering certain things with different strategies. Others tend to confuse with eidetic memory, although it is a more common issue among children than adults (Foer, 2016). However, many people say that photographic memory exists, and it simply isn't understood correctly. It does not work like observing a photograph, after all. Instead, you have to use techniques to remember anything that's already in your memory bank. Nevertheless, now that you have learned a variety of strategies to increase your photographic memory, it is time for you to decide for yourself: does photographic memory exists or not?

Through the basic and advanced techniques that you learned in this book, you should be able to improve your memory. You may not find this to be true immediately; it can also take a bit of time to fully understand and use the ideas naturally. Despite that, through patience and determination, you will be able to overcome any problems and start to upgrade your memory.

Not only did you learn about what memory is, but you also saw the three phases of memory and how the memory process could be interrupted. At the same time, you learned about the different types of memory, with a special focus on photographic memory. Of course, you were able to get an idea of what type of benefits photographic memory will give you because, as many people know, you always want to understand why you should work towards something. The reasons described in this book, such as being able to perform better academically, boosting your confidence, becoming more mindful, and remembering specific information better are some of the reasons why you should build your photographic memory.

Lifestyle improvements are another way to work towards improving your memory as well. In fact, when you can get enough sleep and exercise, creating your own memory palace becomes easier than you think. Along with this, you also know how to create your own mind map and understand how mnemonics work. This is a great start to make sure that you realize both the basic and advanced techniques discussed in this book, from the SEE principle all the way to the emotion-based method.

It is important for you to know that your learning does not stop here. In fact, you can continue to build your memory through my next two books in this series. The second book called *Memory Training* focuses on brain training and memory games. Then, you follow it up with the third one, which will be known as *Memory Improvement.* The latter concentrates on the healthy habits that you can install into your life in order to build your memory. Because this is the first book of the series, though, you want to take your time to understand at least some of the techniques mentioned in the previous chapters.

Furthermore, it is possible that there are some —

such as the car method or the linking technique —
that you will not like just because they don't fit with
your personality. Still, remember that you should
never stop improving your memory. Even if you find
yourself joining a memory competition around the
world, you want to continue to have the best
memory possible. Not only will this help you
remember a variety of information through your life,
but you will also be able to decrease your chances of
developing cognitive disorders, such as Dementia
and Alzheimer's disease.

Your brain is one of the most important parts of your
body. Therefore, you have to do whatever you can in
order to keep it active and healthy. By doing so, you
can accomplish more things, feel more energized,
and improve your mental and physical well-being.

The way I see it, there is nothing negative about
taking at least 15 minutes out of your day to make
sure that you are doing everything to allow your
brain to keep on performing at its best.

Having a developed photographic memory is a very unique skill that, will give you an edge over all the people around you.

UPGRADE YOUR MIND -> zelonimagelli.com

UPGRADE YOUR BUSINESS -> zeloni.eu

EDOARDO ZELONI MAGELLI

MEM●RY TRAINING

Memory Games and Brain Training to
Improve Memory and Prevent Memory
Loss
-
Mental Training for Enhancing Memory
and Concentration and Sharpening
Cognitive Function

EDOARDO
ZELONI MAGELLI

EDOARDO ZELONI MAGELLI

MEM●RY
IMPROVEMENT

The Memory Book to Improve and Increase
Your Brain Power
-
Brain Food and Brain Health Habits to Enhance
Your Memory, Remember More and Forget Less

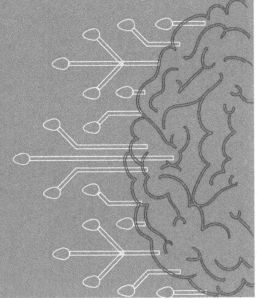

EDOARDO
ZELONI MAGELLI

Bibliographical References

Alban, D. (2018). *36 Proven Ways to Improve Your Memory.* Retrieved from https://bebrainfit.com/improve-memory/

Beasley, N. (2018). *Difference Between Eidetic Memory And Photographic Memory.* Retrieved from https://www.betterhelp.com/advice/memory/difference-between-eidetic-memory-and-photographic-memory/

Boureston, K. (n.d.). *How to Develop a Photographic Memory: The Ultimate Guide.* Retrieved from https://www.mantelligence.com/how-to-develop-a-photographic-memory/

Buzan Tony, Buzan Barry (2018) *Mappe mentali. Come utilizzare il più potente strumento di accesso alle straordinarie capacità del cervello per pensare, creare, studiare, organizzare*

Foer, J. (2016). *Slate's Use of Your Data.* Retrieved from https://slate.com/technology/2006/04/no-one-has-a-photographic-memory.html

Friedersdorf, C. (2014). *What Does it Mean to 'See With the Mind's Eye?'.* Retrieved from https://www.theatlantic.com/health/archive/2014/12/what

-does-it-mean-to-see-with-the-minds-eye/383345/

Improve Your Memory With a Good Night's Sleep. (n.d.).
Retrieved from
https://www.sleepfoundation.org/excessive-sleepiness/per
formance/improve-your-memory-good-nights-sleep

Kubala, J. (2018). *14 Natural Ways to Improve Your
Memory.* Retrieved from
https://www.healthline.com/nutrition/ways-to-improve-
memory

Lerner, K. (n.d.). *Hook Line & Sinker - Secrets to a Great
Memory Hook.* Retrieved from
https://www.topleftdesign.com/blog/2009/11/hook-line-
sinker-secrets-to-a-great-memory-hook/

Mcleod, S. (2013). *Memory, Encoding Storage and
Retrieval.* Retrieved from
https://www.simplypsychology.org/memory.html

Memory Process - encoding, storage, and retrieval.
(n.d.). Retrieved from
http://thepeakperformancecenter.com/educational-
learning/learning/memory/classification-of-memory/
memory-process/

*Memory Techniques - Association, Imagination and
Location.* (n.d.). Retrieved from
https://www.academictips.org/memory/assimloc.html

Method of Loci - Increase Memory Using your Home's

Map. (2011). Retrieved from https://www.mind-expanding-techniques.net/memory-strategies/method-of-loci/

Mind Mapping - How to Mind Map. (n.d.). Retrieved from https://www.mindmapping.com/

Mind Mapping Basics. (n.d.). Retrieved from https://simplemind.eu/how-to-mind-map/basics/

Mohs, R. (n.d.). *Improving Memory: Lifestyle Changes.* Retrieved from https://health.howstuffworks.com/human-body/systems/nervous-system/improving-memory1.htm

Negroni, J. (2019). *How to Memorize More and Faster Than Other People.* Retrieved from https://www.lifehack.org/articles/productivity/how-memorize-things-quicker-than-other-people.html

Pinola, M. (2019). *The Science of Memory: Top 10 Proven Techniques to Remember More and Learn Faster.* Retrieved from https://zapier.com/blog/better-memory/

Qureshi, A., Rizvi, F., Syed, A., Shahid, A., & Manzoor, H. (2014). *The method of loci as a mnemonic device to facilitate learning in endocrinology leads to improvement in student performance as measured by assessments.* Retrieved from https://www.ncbi.nlm.nih.gov/pmc/articles/PMC4056179/

Step 3: Memory Retrieval | Boundless Psychology. (n.d.). Retrieved from https://courses.lumenlearning.com/boundless-psychology/chapter/step-3-memory-retrieval/

The Good And Bad Things. (n.d.). Retrieved from https://photographic-memory-science.weebly.com/the-good-and-bad-things.html

The Journey Technique: – Remembering Long Lists. (n.d.). Retrieved from https://www.mindtools.com/pages/article/newTIM_05.htm

The Study of Human Memory. (n.d.). Retrieved from http://www.human-memory.net/intro_study.html

Types of Memory. (n.d.). Retrieved from https://learn.genetics.utah.edu/content/memory/types/

Types of Memory | Boundless Psychology. (n.d.). Retrieved from https://courses.lumenlearning.com/boundless-psychology/chapter/types-of-memory/

Wik, A. (2011). *How To Remember Anything Forever with Memory Hooks.* Retrieved from https://roadtoepic.com/remember-anything-forever-with-memory-hooks/

CPSIA information can be obtained
at www.ICGtesting.com
Printed in the USA
LVHW081901051120
670660LV00017B/335